What Darwin and Dawkins Didn't Know:

Epigenetics, Symbiosis, Hybridization, Quantum Biology, Topobiology, the Sugar Code, and the Origin of Species

John A. Rush

Amazon/Kindle Publications

Copyright © John A. Rush All rights reserved. No portion of this book, except for brief review, may be reproduced, stored in a retrieval system or transmitted in any form or by any means – electronic, mechanical, photocopying, recording, or otherwise – without written permission from the author. For information contact: jarush43@gmail.com.

Published by Amazon
ISBN: 9798670598743

Table of Contents

Preface v
Chapter 1 – Introduction 1
Chapter 2 – Classical Genetics and Neo-Darwinism 27
Chapter 3 – Epigenetics 53
Chapter 4 – Symbiosis 95
Chapter 5 – Hybridization 117
Chapter 6 – The Quantum World, Consciousness, and Reality 125
Chapter 7 – Hominin Evolution 157
Chapter 8 – Conclusion 181
Bibliography 195
Index 217
Author Profile 229

Preface

Human behavior is highly ordered, highly ritualized. We create narratives and models regarding how the world works in order to satisfy a need for order or meaning in our lives. But sometimes our initial assumptions are wrong and the models we create lead to misunderstanding and following a false path that leads nowhere.

Charles Darwin, in his *On the Origin of Species*, presented a simple model, one that matched his view of the world, and that is the idea of common descent. Looking at the paleontological data one can see that animal forms show common features suggesting gradual change over long periods of time resulting in new species. This was termed *uniformitarianism*. Darwin, however, didn't have a mechanism illustrating how this occurs, but it sounded good and appeared correct. But appearances can be deceiving. Shortly after *On the Origin of Species* was published (1859) sociologists, in the likes of Hubert Spencer, equated biological evolution with social evolution and "social Darwinism" raised its ugly head. This eventually entered the politic with a belief that the most fit should rule. This then led to eugenics, socialism, Nazism, and communism where a small group of elite scholars and politicians should set the rules for everyone else. In colleges and universities there is a great push to indoctrinate students into believing that socialism and communism should be the rule of the day, disregarding the history connected to these elitist systems and the common misery caused by this philosophy. Once Darwin's position was connected to the politic it became an unshakable truth, repeated over and over again.

With the rediscovery of Mendelian genetics and the isolation of DNA in the early part of the Twentieth-Century, and then the discovery of the double helix, nucleotides, and amino acids, Neo-Darwinism was born. In short, the mechanism of change was clear; the process was random and without purpose – life just accidently emerged in tidal pools during the early part of Earth's history. All that had to occur was the alteration of base pairs, or the sequence of base pairs, by random copying errors and other acts of nature over a long period of time, and, bingo, new species are born. Thus, the model, the narrative became the process and, unlike the theory of gravity, was never described or tested using the materialistic science of the day. The model, Random Mutation + Natural Selection + Time = New Species, has never been proven using materialistic science. How (the process) one would convert inorganic matter into organic matter, and then to self-sustaining life has never materialized. How, through random acts, can you create a functioning protein let alone a cell or any complex biological system such as glycolysis or blood clotting? This has never been elucidated in any tangible, demonstrable way by the Neo-Darwinists.

When looked at through a scientific lens, random mutations can't create anything outside of perhaps a point mutation and perhaps a minor phenotypic change. Most mutations, however, are lethal. All genetic coding can be changed – if you follow the rules, some of which we are now beginning to uncover. Random mutations do not follow the rules. In short, when you alter an amino acid, in most cases, other changes have to occur as well or information is removed from the system leading to death, ill health, or nothing happens. One aspect of genetics the Neo-Darwinians didn't contemplate is that cells can think and make decisions. Life forms are not simply at the mercy of the environment and what the environment throws at them. Biological systems, and the cells out of which they are composed, are active participants in their own evolution, making choices and altering coding sequences in response to environmental stressors. What the Neo-Darwinists assumed were random alterations may not be random at all but purposeful alterations, and when the cell alters a coding sequence all the auxiliary systems are altered as well. Moreover, most of these complex alterations occur in the wave form, which would solve the problem connected to the construction of irreducibly complex systems. This, as the reader will appreciate, might also help to explain the Cambrian explosion and the development of many new species in general, especially after catastrophic events. Of course, thinking and choices smack of purpose, and purpose reeks of intelligence, something the Neo-Darwinists actively fear and fervently deny; heaven forbid we live in an intelligent, purpose-directed universe! And what might that purpose be? At a very basic level, life slows entropy.

The following represents what Darwin and Dawkins did not know, and, in the case of Dawkins, pushed aside as irrelevant; this is what I term, Hound Dog Science. Let's look at these "irrelevant" processes as they promise to lead us toward a more scientific, and realistic exploration of the origins of life and the origins of species. I'll let the reader decide if the academic community has, out of political motivation, misrepresented science for over 100 years.

I would like to thank Katie Nemer and Ron Baioni in the production of this manuscript, Hannah at 99designs.com for cover design, and Nick Caya and his associates at Word-2-Kindle for final formatting and editing.

John A. Rush
June 2020

CHAPTER 1
Introduction

CHARLES Darwin's, *On the Origin of Species* (1859), is seen as an attempt to identify the origins of life on this planet and the biological changes that followed. His basic premise is that change occurs gradually over long periods of time leading to new species. Darwin did not, in any detailed way, describe the process of evolution, nor did he ever use the term "mutation," just that change was slow and accumulative. However, he had reservations about his position, especially considering the Cambrian explosion, with different phyla just showing up without precedents. He thought, however, geologists would uncover the precedents or intermediate forms in time; he had faith. He also stated:

> If it could be demonstrated that any complex organ existed which could not possibly have been formed by numerous, successive, slight modifications, my theory would absolutely break down (Darwin 1988 [orig. 1859]: 154)

The concept of "irreducible complexity" (Behe 1996) shows that most biological systems (the structure of a cell membrane, citric acid cycle, glycolysis, blood clotting, immune functions, on and on) could not have "formed by numerous, successive, slight modifications." To date, no one has proved otherwise, although there are hollow claims that some alleged irreducible complex systems "are entirely reducible" (see Chapter 8). We term the Twentieth-Century upgrade Neo-Darwinism and it is a good example of where a little knowledge is misleading and a danger to scientific method. The double helix and the identifiable nucleotide bases cinched it for the Neo-Darwinists. The process of evolution involved alteration of base pairs or their sequence. The formula for Neo-Darwinism is: Random Mutations + Natural Selection + Time = New Species. The mechanism for developing new species is random, spontaneous mutations acted upon by natural selection. In short, random mutations plus natural selection *is* the process. In fact, Random Mutations + Natural Selection + Time = New Species is an idea, a hypothesis. The process would be explaining, in detail, using materialistic methodology (i.e., math, chemistry, and physics), how random mutations can lead to self-sustaining life forms and new species. Surprisingly (or not so surprisingly) Neo-Darwinists have not, to date, been able to prove their case. Yes, there has been research on fruit flies and bacteria, by inflicting mutations through radiation, for example, but instead of developing new species they have developed monsters (fruit flies with extra limbs sticking out of their heads, etc.) or weakened the insect or strain of bacteria (see Behe 2007). Those that cling to this formula, random mutations, natural selection, and time leading to new species (bitter clingers to use a political jab) do so on faith, just as many have faith in the existence of gods and demons. The Neo-Darwinists leave it up to others to prove their case; the proof does not lie in the idea, hypothesis, or word, any more than the word "gravity" explains why an apple falls from a tree. The word gravity is compressed data that includes material science (the math and physics) to explain the effects. Random mutation is also compressed data but with no support from material science.

For over 50 years a steady influx of research data from many academic fields, including cell biology, genetics, and mathematics/statistics, confirms the dominant theory of biological evolution, Neo-Darwinism, is inadequate to explain the sudden appearance of numerous separate phyla at the beginning of the Cambrian (550-450 MYA), let alone the origins of life itself. Nor is the hypothesis adequate to explain the complexity of DNA and all the intricate systems that support life leading to new species. Neo-Darwinism is a faith-based model; one has to have faith that the model is correct without scientific evidence to back it up. Why this model has been the dominant explanation for the past 100 years, despite solid evidence to the contrary, will be exposed in Chapter 8.

Science is a process of discovery and one of the best and recent examples of this is cosmology and the discoveries revealed through advanced telescope and sensing technology, space probes, and so on. Many old models were replaced almost overnight with new data that could not be ignored. For cosmologists these days, data comes in at such a clip there is no time to establish dogma, ideas and interpretation, yes, but intrenched dogma, no. Many of the cosmologists, for example presented on the TV science program, *How the Universe Works*, still present the random mutation, Neo-Darwinists' model as the way the universe and life forms came about. Within their own discipline, however, cosmologists are open to new interpretations.

Paleoanthropology had its dogma turned around when Piltdown man, through fluorine dating, was revealed to be a hoax, allowing legitimate finds, the Taung infant (*Australopithecus africanus*) and *Homo erectus* from Java, to take their proper place within what we think we know about our origins. Lucy (*Australopithecus afarensis* c. 3.2MYA – Million Years ago), Johanson's 1973-74 find, was considered the base of the family tree, dogma, for many years until *Orrorin tugenensis* (c. 6MYA), also a biped, showed up creating doubt. For the past two decades more bones of our supposed ancestors emerged in various places (e.g. Flores, Indonesia – *Homo floresiensis* c. 90-12KYA; Rising Star Cave, South Africa, *Homo naledi* c. 250KYA – I have much more to say about hominin evolution in Chapter 7), and through genomic research paleoanthropological dogma is falling apart at the seams. Yes, there are some who consider Lucy the base of an imagined family tree, which is highly unlikely (see Chapter 7); these are the bitter clingers in paleoanthropology and biological anthropology.

For Neo-Darwinists and their view of evolution, we have a different situation. By the early 1950s the model had become so intrenched that any critique was quickly pushed aside. This faith-based position had become political showing a weak faith in the model, so weak in fact they have to demonize others offering different perspectives. And, heaven forbid, any opposition could cause the rewriting of textbooks, possibly undermining years of research, and perhaps damaging reputations of the bitter clingers in that arena. New technologies have allowed researchers to understand cellular biology and RNA and DNA and the effects of hybridization, symbiosis, and epigenetics, in great detail, far greater than Darwin (and apparently Dawkins 1996, 1999, 2019) could or can imagine. In fact, if Darwin had access to our current information he would never have written *On the Origin of Species*, or at least he would expand his model to include what we have discovered in only the last 50 years. He might have become a Lamarckian.

Unlike cosmology, many in the academic community ignore new concepts regarding evolutionary processes and thus science takes second place to politics. A good scientist considers all the possibilities, without political prejudice, and tests to confirm or discard the differing options. From a scientific standpoint, Neo-Darwinism is the major model, or we should look elsewhere. Neo-Darwinism is not testable; and, just like gods and demons, we can only accept it on faith, and that's not good science. In fact, that's not science at all. There is such insecurity in their faith that Neo-Darwinism supporters often angrily talk down the opposition, ruin reputations, shut off such discussions during staff meetings, and remove "infidels" from university positions. This is politics and not science, just as "global warming" has become a political issue with no rational debates as to cause. To contradict the concept of "man-make" global warming invites harsh criticism. We are in an interglacial period and we're supposed to be warming up, and in 50 KY (thousand years) or so we will be back into another ice age. There have been 6 or 7 major ice ages/interglacial periods during the past 3 million years with several less global (e.g., Younger Dryas–10,900 BCE). Staying with the Younger Dryas, it is interesting to note that after 1200 years (by c. 9,600 BCE) the mini-Ice age ended, after which there was a rapid warming period almost as if "nature" was trying to make up for lost time, with glacial melt occurring rapidly to make up for the lack of warming for 1200 years. What causes these interglacial periods? Sun spots, supernova explosions, pole reversals, volcanos, too much oxygen in the atmosphere, methane extruding from animal life and the sea bed, gasses released in the manufacturing process, humans breathing and politicians talking, meteors and comets, atom and hydrogen bomb testing, ants, farting cows, on and on. Once politics enters science, the data presented as "scientific" cannot be trusted. The scientists who wrap themselves in the political blanket, in my opinion, can't be trusted either. Neo-Darwinism is no exception.

Before reading further, my focus in not on Intelligent Design. Intelligent Design is just a few words that originally meant "we don't know; there appears to be some intelligence behind this," and can stall the process of investigation by frightening Neo-Darwinists and creating division in the ranks. Because organized religion hijacked ID, I prefer the term *Congruent Construction* (see Chapter 8). If there is an intelligence behind the order of the universe or the origins of life forms, this "intelligence" has to play by the rules, whatever their origin; this energy can't say ABRACADABRA, with a puff of smoke, and have life jump out of a hat. "In the beginning was the Word and the Word was with God and the Word was God" (John 1:1) is a good example of this thinking. A process would have to be followed. I'm interested in the processes, the rules, how this all works, how you get from inorganic material to organic life forms, and so on. Invoking the supernatural or magical helpers, like random mutations and natural selection, does not satisfy me at all and only leads to dogma and misunderstanding. I have "faith" that some energy informs this universe, but I do not understand what it is. None of the "religious" traditions (Hinduism, Buddhism, Judaism, Christianity, Islam, etc.) have a handle on this energy. In fact, these traditions (especially

Judaism, Christianity, and Islam) believe that morality stems from their particular religious/political tradition. This is untrue, morality is a social necessity and begins in the family long before individuals have any inkling of the supernatural world. In fact, Judaism, Christianity, and Islam, and especially Islam in more recent times, are responsible for the murders of millions of innocent people just because they did not submit to their demonic code, God the Father for Christianity and Hiliah or Allah, the Moon God of War, for Islam (see Rush 2008). The Catholic Church was also behind the killing of thousands and thousands of innocent felines (cats), labeling them as instruments of the Devil and able to do His bidding; this attitude toward (fear of) cats has lasted into modern times.

Judaism, Christianity, and Islam have no special place when it comes to morality and human decency. The anthropomorphic gods supporting these traditions are demons. Demons ask for absolute submission, to kill when required, and mental and emotional enslavement. Those traditions that applaud the Intelligent Design position, thinking the reference is to their personal god, help to contaminate the investigation. Their models/beliefs, like cold fusion and Neo-Darwinism, should be placed on the shelf, in plain view, awaiting further discoveries. If there is a god that created all this, one would have to agree that the intelligence is far beyond anything we can image, and I have great difficulty believing that this intelligence, this energy, is so insecure that it needs to be worshiped!

Although this book represents a short review of major problems with Neo-Darwinism uncovered to date, my main thrust rests with the identifiable and testable processes of biological evolution or change. In short, Neo-Darwinism presents an inadequate, untestable model, and in the following chapters I will present an alternative.

The Four Big Questions

1) What is the origin of the universe (cosmologists suggest many possibilities)?

2) How did self-replicating life forms emerge from inorganic material? Or, what is the origin of life (biologists offer numerous possibilities)?
3) What is the nature of evolutionary change (chemically, physically, mathematically), how does this all work from a materialistic position, and are there parts missing from our current model(s)?
4) What is the evolutionary basis for an evolutionary process? Where did the patterns and codes come from leading to enzymes, RNA, and DNA? Are random mutations all that is necessary to build the universe and the life forms within?

Origins of the Universe

Let's begin with origins of the universe. There is no testable, verifiable model regarding the origin of the universe. The most popular position is the "Big Bang" theory, originally suggested by a Belgian priest named Georges Lemaître in the 1920s, when he theorized the universe began from a single "primordial atom." Edwin Hubble's research amplified this idea. Photographing galaxies, Hubble noticed "red shifting," caused by what we call the Doppler Shift, showing that these galaxies are moving away from us at great speeds. Sound waves are a good example. Standing by the side of the road, a moving vehicle approaching you is compressing the sound generated by air friction, tires pounding on the road, and engine noise; this produces a high-pitched sound. As the vehicle moves past and away from you, the sound waves flatten out (decompress, lose energy) producing a lower-pitched sound. Light waves act in a similar manner. Light moving away from you shifts toward the red portion of the light spectrum, while light moving toward you, like the Andromeda Galaxy (Messier 31), shifts toward the blue portion. This suggested to Hubble and other researchers that there must have been an "explosion" of sorts emanating from a point of origin, thus Lemaître's "Big Bang." The problem with the explosion metaphor is the universe, yes, is lumpy because of the formation of galaxies, but it is very uniform. Explosions are not like that. The suggested time for this event was originally reckoned to be greater than 8 BYA; our current position is the universe is about 13.8 BYA. We are talking about some type of energy, densely packed, existing somewhere at a particular point in time, although space and time didn't exist. For some unknown reason this energy—obviously highly organized and operating with a set of defined rules (some of which we understand) encapsulated in physics, mathematics, and chemistry, is released, rapidly inflates, moves out from a center, creating space and time and a uniform universe in the process (although some researchers see space and time as mental reference points for fitness rather than tangible "things"; see Hoffman 2019).

Thus, we go from a "singularity" to a Big Bang, or release of energy (plasma/quantum wave form[s]) and the formation of a dualistic universe containing two general types of energy, plus and minus, the one type of energy tugging at or repelling the other. Then we have cooling, the formation of hydrogen (particles), then clumping/gravity/heat, star formation, more heat/pressure producing helium, etc.

Questions: What is a singularity? Where did this "singularity" come from? Was this a random event? Where did all the energy involved in this initial event come from? Would not hydrogen have to be observed or measured before it can become a particle to compress into helium? Is there an end to the universe? Math and physics help to build these models, but we may never understand the origins of the universe, where it came from or how it came into existence. The origin of the universe and the origin of life are intimately connected. After all, we are a product of star dust.

Further, the Big Bang (I'll continue to use this analogy for convenience) released a tremendous amount of energy—again, where did this energy come from? Eventually hydrogen emerges with this release of energy—is this now a random event? Why hydrogen—why not gold? The way subatomic particles and elements combine through cooling, clumping, heating, and effects of gravity leading to star formation, is this process random? Gravity is not something that just randomly happens; neither is the clumping behavior of hydrogen atoms. If this process of star and galaxy formation were random, either nothing would form or there would be much more variability in the constituents of stars. If the "Big Bang" was truly a cluster of random events, in my opinion, there would be no universe as we know it, no galaxies, stars, planets, and the ancient alien hunters would be out of business.

The Big Bang, the cataclysmic event that many believe brought the universe into existence some 13.8 BYA, was, according to quantum physicists, triggered by a random fluctuation in what we call the "wave form," with particles popping into and out of existence. Yes, they can, but scientists do not truly know what is happening in the wave form, and whatever is happening it isn't random, another term used when we don't see the bigger picture. Cosmologists resort to metaphors to express ideas about possibilities—few accept these metaphors as absolute truth. But while some physicists believe this bang, pop, or crackle and the resulting "inflation" of the universe was a singular, unique event, others say there could have been many such events — leading to multiple universes. One argument for a multiverse (multiple universes) develops from string theory. Matter is composed of unimaginably small, vibrating strings or loops of energy—not particles. Physicists hoped string theory might be a "theory of everything," or a group of equations that explain why our universe has the exact properties that it does. For example, why is the mass of a proton 1836.15 times greater than that of the electron? No one has a good explanation. But instead of a single solution, string theory's equations seem to have an unbelievable number of solutions (perhaps 10^{500} — that's 1 followed by 500 zeros or one-hundred-cenquinsexagintillion!). To some scientists, each of these solutions describes a different universe, each with its own physical properties.

Another claim for the multiverse comes from quantum theory. Quantum theory has proven successful at describing matter on the smallest scale. However, quantum theory leads to several possibilities that defy common sense. The "many worlds" interpretation of quantum theory suggests that the universe splits in two each time there's a so-called "quantum event" (see Carroll 2019 for a clear presentation of the multiverse position; also see Vaas 2017 for a critique of multiverse scenarios). In the strange world of quantum theory, a radioactive particle decays and doesn't decay during any given period — and each result exists in a separate universe. With such quantum events happening more or less continuously, the argument goes, the number of universes keeps increasing. We can test none of this outside of computer models, models constructed by an intelligence using today's knowledge and models that are admittedly incomplete. I will return to multiverses in Chapter 6.

Teleological (Considering Purpose Rather than Cause)

Besides the laws governing the behavior and order of elements, there are also principles of self-organization or the development of complexity over time not explained by those elemental laws (math, physics, and chemistry). Our models are incomplete (Kauffman 2019). Teleology searches for a purpose, i.e., what is the purpose of life or what is the purpose of a nervous system that places values on objects and behaviors? How do ideas about "right" and "wrong" emerge from our nervous system? Why would we humans have a brain capable of putting a person on the Moon at least 70 KYA (Thousand Years Ago)? If there is purpose, then that smacks of intelligence; the Neo-Darwinists generate a great deal of fear over this prospect.

God, Grand Craftsman, Designer

Then we have the easy way out. A specific, anthropomorphic "consciousness" set the rules of the universe and put everything in motion. This consciousness is called God, Grand Craftsman, Brahma,

Tiamat, Pan Gu, etc. Calling on a god to explain origins does not inform how the universe and organic life emerged.

Perhaps we live in a computer Matrix, another easy way out. How about the Ancient Alien hypothesis, that aliens from other worlds or dimensions designed us primarily to work as slaves mining gold (Sitchen 2014)! I don't know if any of these proposals are correct— but I do not discount them, for to do so is not good science. So what if there is intelligence, consciousness, or God in back of this? Should this stop us puny humans from discovering how all we experience came together? Those who vehemently oppose such considerations appear to have little faith in their own models, or perhaps a great fear of these possibilities, of intelligence or God.

Origins of Life on Planet Earth

It is difficult to place a date on the beginning of life in the universe. It would have happened once necessary chemicals were available, i.e., hydrogen, helium, lithium, carbon, oxygen, sulfur, phosphorus, and so on, and as they form in stars, the various star formations would probably give us a relative date. The first stars apparently formed about 300 million years after the Big Bang and I would suspect that by at least 10 BYA (Billion Years Ago) (if the universe is 13.8 billion years old) we would have neutron stars, which may be responsible for producing the heavy metals. However, many scientists suggest that planet Earth formed in our galaxy around 4.4 BYA, with some thinking life first began on planet Earth 4-3.5 BYA. With the Earth proper forming 4.4 BYA, as we will see, there is just not enough time for life forms to emerge randomly. Life, or rather the generic rules for constructing the cell and life forms that we know as bacteria, were available in the wave form that spawned the universe, and most likely has been available since the beginning of time and space and not peculiar to planet Earth. Let's consider models that reflect the Earth-centric view of life's origins (see Fry 2000 for a history of this quest for life's origins).

Carbon-based Life Forms

Life on planet Earth is carbon based. Science fiction writers, however, imagine beings made of silicone or perhaps aluminum. Before I get to the problems of using silicone or other materials besides carbon let me first state that all life forms on this planet use redux reactions (reduction-oxidation) for moving electrons from one molecule to another. One molecule gives up an electron, and another molecule accepts an electron; it reduces the substance that receives the electron. For example, oxygen O_2 during oxidation picks up two electrons and two protons, balancing the charge and reducing O_2 to water (H_2O). In order for redux to occur, you need to put a certain amount of energy into the reaction. As the reaction proceeds, it releases heat increasing entropy. We use energy to slow down the loss of energy, and this system has to be cost effective. Now let's consider carbon vs silicone as a building block of life. Lane comments (2015: 78):

> Could life have used something other than carbon? No doubt it is conceivable. We are familiar with robots made from metal and silicone, so what is special about carbon? Quite a lot, in fact. Each carbon atom can form four strong bonds, much stronger than the bonds formed by its chemical neighbor silicon. These bonds allow an extraordinary variety of long-chain molecules, notably proteins, lipids, sugars and DNA. Silicone can't manage anything like this wealth of chemistry. Also, there are no gaseous silicone oxides to compare with carbon dioxide. I imagine CO_2 as a kind of a Lego brick. It can be plucked from the air and added one carbon at a time on to the molecules. Silicone oxides in contrast . . . well, you try building with sand. Silicone or other elements might be amenable to use by a higher intelligence such as ourselves, but it is hard to see how life could have bootstrapped itself from the bottom up using silicone.

> That's not to say that silicone-based life couldn't possibly evolve in an infinite universe, who could say; but as a matter of

probability and predictability . . . that seems overwhelmingly less likely. Apart from being much better, carbon is also much more abundant across the universe. To a first approximation, then, life should be carbon based. I suspect that carbon-based life is the rule in the universe. It is more versatile and energy efficient when combining with other molecules, and even if silicone is the favorite in some other universe it is of little concern for us; let's figure out one universe at a time.

Primordial Soup Model

Once Upon a Time . . . the waters and atmosphere on planet Earth were rich in chemicals, including carbon dioxide, carbon monoxide, hydrogen cyanide, ammonia, and so on, or so goes the narrative.

Fry (2000: 114) comments:

The question of the composition of the early atmosphere is still hotly debated. Several crucial factors seem to tip the scale toward a non-reducing atmosphere, but the picture is not at all clear, and there are indications that even though the major component was carbon dioxide, methane was present. Some scientists still insist on a reducing atmosphere, while others favor a nearly neutral oxidation state (Sagan and Cbyba 1997: 1217).

This debate is crucial to the origin-of-life question, since the postulation of a non-reducing atmosphere makes it difficult to explain the generation of simple organic molecules and their chemical evolution to more complex ones. In short, if there was little or no free oxygen then redox reaction would bring forth amino acids, and so on; if free oxygen was present, then these complex chemicals cannot form. Methane and amino acids are precisely the chemical signatures cosmologists search for when considering the probability of life existing, for example, on one of Saturn's moons, Enceladus, once thought to be a simple ball of ice.

The Earth represents a closed system and electrical discharges from lightning storms, within the thought-to-be components for Earth's early atmosphere, would provide the energy for the production of amino acids. Eventually amino acids formed proteins, proteins formed enzymes, then RNA then DNA—all randomly.

Problems

Scientists aren't sure if the early Earth atmosphere was reducing;

There is no explanation given as to how left-handed amino acids developed independently from right-handed amino acids and how they developed into enzymes, etc.;

UV radiation would destroy any life forms close to the surface in tidal pools;

Not enough energy available through lightning strikes to fuel reactions;

The moon was much, much closer, orbiting 15,000–20,000 miles away, as opposed to the current average distance of 238,000 miles.

I'll return to these problems one by one.

The Miller-Urey Experiment

In 1952 Stanley Miller, with his advisor Harold Urey, put together an experiment designed to test earlier beliefs that, by looking at the atmospheric chemistry of the early Earth, one might determine if these chemicals could bring forth life forms. For their experiment, Miller created a closed system that included water, chemicals he assumed were in the atmosphere of early earth (ammonia, hydrogen cyanide, carbon monoxide, carbon dioxide, and so on), two energy sources, one a heat source to warm the water, and the other an electrode through which the cooled gasses would pass. After a very short time period Miller noticed that the color of the water had changed, and, upon examination, could detect amino acids (analysis decades later revealed about 20), tars, and other organic, gooey materials. This experiment, reported faithfully in modern texts, was (and still is in some academic quarters) hailed as proof the development of life was an issue of chemistry. But is this true? What Miller's experiment shows is that amino acids are easy to create in nature—we can detect this material in galactic clouds, but that's it. No self-replicating life forms emerged from Miller's experiment. Moreover, all the amino acids found in Miller's experiment (and similar investigations) are racemic, that is, they are composed of dextrorotatory (right) and levorotatory (left) forms of a compound in equal proportion. All life forms on planet Earth use *left-handed* amino acids (L- or levorotatory) and right-handed nucleic acids, which suggests (and only suggests) all life on planet Earth originated from a single ancestral form using only left-handed amino acids. And if you think about this for a moment it also suggests the origins of life in the universe is very, very ancient and didn't simply evolve here on planet Earth. I am not, however, belittling Miller's experiment; we need to test theories so we can say "yes" or "no" to a specific hypothesis. Neo-Darwinists have never tested their position and what they point to as proof, for example, bacteria "evolving" to deal with antibiotics, is not evolution; no new species developed, the fitness of the bacteria is compromised with these random alterations, and information in the system is lost, not gained—you need new information to build new species.

The Miller-Urey experiment was prompted by the scientific imaginings of little ponds, by the sea, that can trap and concentrate organic materials that ultimately turn into the building blocks of life, and eventually self-replicating life, as we know it, crawled out. The narrative sounds good, and it is a start in the investigation, but there are lots of problems. First, there is no way, in a pond or eddy, to trap information so it will stay associated. There is no way to create a cell. Besides that, cells cannot function unless they are complete with cell pores and the machinery and codes to produce RNA, DNA or any of the support systems—a functioning cell has to show up all at once. The cell is an irreducibly complex system and to think it could come together in increments, one piece at a time via random mutations, is unlikely—at least in a classical physical world. And, again, according to the Neo-Darwinists (Dawkins 2006) this occurred randomly over vast periods of time. But there are other issues.

Second, random activities do not produce patterns or codes—RNA and DNA are codes. Yes, I know what seems chaotic, like weather patterns, is only chaotic because we don't know all the parameters. There can be no such thing as a random, chaotic system, as "system" implies process and rules.

Third, on planet Earth this "evolution" from inorganic to life-sustaining organic material would have occurred during a very short window of time, between 4.4 BYA and 4-3.5 BYA. Statistically, this could not have happened, randomly, in such a short time (Axe 2004, 2017), that is, in under a billion years. For the base pairs there are 16 possible combinations in a short stretch of DNA: ATCG, TACG, ATGC, AGTC, CATG, GTCA, CAGT, CTAG, etc., that build 20 amino acids used by life forms on this planet, along with 44 redundant combinations that can stand in place of the 20 when necessary. I'll discuss redundancy in Chapter 2. Let's consider an 8 letter pass code with both upper- and lower-case letters: Est77991. There are 10^{14} (that's 1 with 14 zeros) or 100 trillion combinations of these letters and numbers; the human genome contains 3 billion base pairs or 10^{38} code combinations (Marshall 2017:35). The time factor alone to create randomly the necessary base pair sequences shows that life-coding organic chemicals most likely came in on comets or meteors, seeding the earth. Time factors also suggest the possibility (the generic rules) of life forms and consciousness existed at the time the universe came into existence or shortly after (in cosmological terms) c. 13.8 BYA. More detailed math is presented in Chapter 2. A key word here is "consciousness," some thought process putting this universe together, a concept vigorously denied by the Neo-Darwinists. As we will see, it is difficult (impossible) to push the idea of a conscious universe aside, especially if we desire to continue to call ourselves scientists. Scientists do not exclude any possibility; scientists stay open to all possibilities—pseudo-scientists do not and insist on forcing faith-based models on everyone else showing they really have little faith in their exulted position.

Fourth, another problem—UV radiation could have provided an energy source but likely (along with cosmic rays) would destroy any life forms, in any stage of development, close to the surface in tidal pools.

Fifth, there isn't enough energy available through lightning strikes to fuel reactions. It would require four bolts of lightning per second, per square kilometer of ocean—not enough electrons in each bolt of lightning to act as an energy source (Lane 2015:92).

Sixth, the moon was much closer, orbiting 15,000–20,000 miles away, as opposed to the current average distance of 238,000 miles. Tidal waves perhaps a mile high (tsunamis even higher) would have raced inland for miles, as land masses on planet Earth were not the same as today. The first supercontinent, formed during the Proterozoic Era, about 2.5 BYA, would have been a relative flat landscape. Waves would wash away any tidal pools or small ponds. A day was 12 hours long—no time at all for any permanent life to form and thrive in such an environment. To add an interesting perspective, according to Takanori Sasaki, from the Kyoto University, the first evidence of life on the planet (where ever or however it formed) 4-3.5 BYA, happened when the day lasted 12 hours. The emergence of photosynthesis, 2.5 BYA, happened when the day lasted 18 hours, 1.7 BYA the day was 21 hours long and the eukaryotic cells emerged. Multicellular life began when the day lasted 23 hours, 1.2 BYA, and the first human ancestors arose 4 MYA (more likely, 8 MYA), when the day was close to 24 hours long. At the very least, it reminds us that solar energy sources have varied over the past 4.4 billion years.

Deep Ocean Vent Model

Black smokers in the Pacific, discovered in 1977, support surprising sea life miles deep in the ocean. These smokers, however, are very hot (250-400 degrees ^{c}) and exist within an acid environment. The vents are hollow, with no place for encapsulating lipids (membranes) or "capturing" amino acids. Thus, we encounter a similar situation as with our primordial soup model—without containment the chemicals simply wash away.

The hydrothermal vents in the Mid-Atlantic, found in 2010, are alkaline environments. They are much cooler (60–90 degrees ^{c}) and the internal structure is like a sponge allowing lipids to concentrate and form bubbles. Bubbles have size limitations and will form a dumbbell shape and then half in two (self-replicate). Once again, the problem is time to create randomly organic, self-replicating organisms. Also, a cell can't function without all its parts and irreducible complexity rears its ugly head, unless something non-random comes into play, something perhaps occurring at the quantum level. None of the above models can answer the larger question: how do inorganic and organic molecules "know" how to come together to produce self-replicating life forms? The rules are not random; only in intelligently constructed computer models do "rules" produce random events—but they are only random because of rules. The stark reality is, we don't know how life formed on planet Earth nor do we know for sure if life began here or somewhere else in the universe. We have to keep in mind our tendency to support an Earth-centered view of the universe, which is naïve. Considering the statistical probabilities, if talking random creation, life forms are ancient and seeded the Earth, just as they have seeded other worlds. Statistically life forms could have been around for 10 billion years or more. This is an opinion, not a fact, but if "life" is imbedded in the fabric of the universe why wait 9 billion years and then just show up on planet Earth? But there is another consideration. We do not exactly know the conditions existing during the early "evolution" of the universe. There may be factors we cannot measure or detect, existing in the wave form, that code for life in the early phases of development. These would be *generic* codes (there are no predestined life forms outside of the initial generic coding), including instructions for the first viruses and bacteria. All other life forms could have evolved from these earlier, generic models through epigenetics, symbiosis, hybridization, quantum biology, and topobiology, and rapidly, especially with thinking and decision making occurring at the cellular and quantum levels. Neo-Darwinists have left the concept of cells thinking and responding to environment stressors out of their model, and this possibility could account for the Cambrian explosion with the production of many phylum and other examples of rapid speciation. The thinking and decision making we possess is a subset of thought or decision making at the quantum level; we might not be smart enough to understand "thought" at that level.

Further, if rules are generic as suggested above, then there is no predetermination, planned out destiny, or specific purpose. What we see in the paleontological record are bones and organs of fish, c. 420 MYA (Late Silurian Period), repositioned or repurposed in later animals—birds, reptiles, and mammals for use/efficiency in new environments (see Shubin 2020). Repositioning organs and bones can have unintended consequences; I will return to unintended design in Chapter 8.

Life is an opportunity to slow down death or entropy and can take many paths; slowing entropy may be the purpose of life. Life forms consume and store energy and thus slow down the inevitable process of entropy or decay. This is the Second Law of Thermodynamics; you can slow down entropy (loss of energy) but you can't stop it, and therefore perpetual motion machines are impossible—at least in this universe. There are lots of models for life's origins but no majority opinion. In science there is no such thing as consensus—something either is, or isn't. When the word "consensus" is used it means we don't know for sure. We know that something is missing in our models. My purpose here is to show that Neo-Darwinism does not help us understand the creation of the universe, let alone the origins of life. Slow, incremental random mutations cannot possibly explain irreducibly complex biological systems. Here is a quick review of the chapters to follow.

Chapter 2–Classical Genetics/Neo-Darwinism

Briefly, Neo-Darwinism, the dominant view of biological evolution, is non-testable and remains a faith-based model. It does not inform as to the origin(s) of life, and cannot show through scientific testing that random mutations can produce irreducibly complex systems like glycolysis, the citrus acid cycle, electron transfer, blood clotting, and so on. In Neo-Darwinism, the genetic code is passive. Research, however, shows that cells and thus genetic materials/support systems are active participants in the evolutionary process rather than passive recipients as suggested with random mutations. Neo-Darwinian dogma is: Random Mutation coupled with Natural Selection plus Time equals New Species, and that "Biological evolution can only occur through mutation or altering sequences of base pairs." Often included is genetic drift (a different mechanism, which may not be random), but results in alterations of base pairs. The concept of random mutations leading to new species is untestable and is likely the *LEAST* important of the evolutionary processes, although random mutations might lead to a new phenotype, usually less fit because information is lost (Behe 2019), but not the development of new species.

Chapter 3–Epigenetics

This is the addition or subtraction of methyl, acetyl, etc., units from DNA/Histones (during gestation and life experiences) enhancing or decreasing expression of the coding sequence(s). Passed on to the next generation they can lead to rapid phenotypic changes and eventual speciation. This shows "thinking" and "decision making" at the cellular/genetic level. Cells being able to "think" and make choices may be a missing part of the origins-of-new-species model.

I also consider the Sugar Code and the part monosaccharides play in cellular communication. Eukaryotes are much more complicated than prokaryotes and require a communication potential way beyond that of nucleotides and amino acids; this is accomplished by various monosaccharides. These combine with lipids (glycolipids) and proteins (glycoproteins) instructing cells to perform various functions from repair to immune activation. There are many interrelated coding systems necessary for maintaining life and I have difficulty conceiving how all this came together randomly.

Chapter 4–Symbiosis

Bacteria and viruses have combined with plants and animals, adding new information leading to rapid speciation. Viruses have probably been there from the beginning, and that beginning of life might extend back 10 billion years or more. Make note of the symbiotic relationship the reader has with gut bacteria, viruses, and so on, not to mention bacteria and mites that live on your skin. For example, there are two types of microscopic eyelash and hair follicle mites, *Demodex folliculorum* and *Demodex brevis*, which feed on dead skin. I'm not sure what we get for supplying mites with dead skin—perhaps more radiant complexions.

Numerous viruses have attached themselves to *HOX* coding sequences (ancient coding sequences) and appear to be very important in the evolution of species. A good example of this is the viral influence on the development of the placenta which occurred around 80 million years ago. We go from being Theria (marsupials) to Eutheria (modern mammals), one species to another, with the help of a virus in a very short time, perhaps one generation.

Chapter 5–Hybridization

Separate species (or subspecies), or species separated for long periods of time (up to 2 MY), that can interbreed and produce fertile hybrids, often have offspring with advantages over parental species. This can lead to rapid speciation. Interbreeding between *Home erectus*, *Homo neanderthalensis*, *Denisovans*, and modern humans is a good example of this. All are variations on a theme, all could interbreed with viable offspring, but it is their hybrid, us, that survived. Possibly inbreeding and accumulated deleterious mutations among erectus-types, Neanderthals, and Denisovans are part of the answer why we, the hybrid, survived and they did not.

Chapter 6–Quantum Evolution

The quantum level appears chaotic, with all the constituents "whirling around in a cloud," or so goes the analogy, the narrative. There is an order here. Thinking/choices are likely made at the quantum level, altering gene expression. Such "thinking" is suggested by the development of an immune system which distinguishes between "self" and "other." This has to coincide with life eating life, or with parasitic entities—probably viruses, or why would we have an immune system in the first place? Where there is life, there is death. This "thinking" most likely reaches back into the emergence of the universe over 13 BYA and may account for new species emerging after catastrophic events, where coding sequences go dormant and during dormancy build new coding sequences. This can happen incrementally because the sequence is not expressed until needed, that is, after the catastrophic event, and that could involve thousands and thousands of years.

Chapter 7 – Hominin Evolution

The evolution of hominins is a hotly debated area of research. At this moment in history, we know little about our prehistoric ancestry. We assume that our lineage began with some erectus-type, perhaps *Homo habilis*, which then "developed" into Neanderthals, Denisovans, and modern humans. Genomics has helped to put some pieces together and in doing so many of our assumption went down the drain. There are three major problems addressed in this chapter—interpretation of fossil remains, dating techniques, and the origins of bipedality. For example, it was once thought Lucy, Johanson's *Australopithecus afarensis*, was the base of the family tree—this is highly unlikely. Tim White's *Ardipithecus ramidus* described as a biped is highly unlikely.

Assumptions made about K-Ar, C-14, and U-238 dating have turned out to be incorrect, and, with many testing procedures used with sites, usually giving different dates, the researchers can choose the date that matches original assumptions. Genomic dating likewise runs on assumptions regarding how many mutations occur per generation.

Paleoanthropologists are especially interested in bipedalism, when and why; the field of topobiology may be useful in solving this puzzle. Topobiology is the study of how contiguous tissues/cells communicate in terms of, for example, sculpting joint surfaces showing that some phenotypic variations emerge, usually to fine-tune a system, through repetitive body movements over time. Can these modifications be passed on to future generations? As there are levels of communication between cells including germ cells, some of Jean-Baptiste Lamarck's examples, once thought extreme, might have some validity.

Chapter 8 – Conclusion

In this chapter I enter the nasty, distasteful realm of politics and politicians, small groups of people with power who think they know what is best for everyone else and thus they go about corrupting science for their own personal ends. Basically, Neo-Darwinism has held on because powerful people, like popes and tyrants of the past, have decided "truth," and you are discouraged from going beyond the bounds of the current myth, that is, randomness created the universe and everything in it. Humans are a small group animal, and our perceived loyalties are mainly to members of these small groups. Yes, we love our country, our states, and the neighborhood, but our family, and our extended family, i.e., friends, and workmates, they are our small group reference. This is where major problems develop, and it is in small groups we decide the next course of action. Contrary to the Marxist position, individuals are the ones who plan and then sell these ideas to others, and in almost every case, decisions made represent the needs of these small groups and members within. Moreover, humans are not like ants in an ant hill. We are a small group animal and that is where our loyalties reside. Again, the major frame of reference is toward one's small group. As an example, we elect a member to the House of Representatives because of his or her promise to work for the community. This individual gets to Washington and immediately realizes that to get anything done for the community one has to "play ball" with others, with their different needs for their community. This leads to "pay for play," and right away the individual is swept into political corruption and protecting one's derriere—if you want reelection. All large systems—name any country in the world, any college or university system, police department, hospital—they all become corrupt, not because that is the nature of the human animal, but because of our small group nature and the rights and obligations that evolve as a member of small groups. In recent decades there has been a push for a one world government, as evidenced by the United Nations, World Health Organization, and so on. One way to get there is through military conquest, but after such an event with current weaponry there likely would be few people to govern. Another way to get to a one world government is to convince people that there are superior people who should rule the world, the philosopher kings, as Plato described it. But, once the individuals get into power, the inevitable occurs—they become corrupt and look after their own needs and the needs of their small group network. So, what does this have to do with Neo-Darwinism? It is a lot easier to control people if we all have the same beliefs, and there is one belief in particular that gets in the way—the belief in a superior intelligence, one that pervades the universe. Intelligence, instead, belongs to the rulers, those on top, and there can be nothing on top of that. Remember, Lenin (Marxist Communism) banned religion because there was no one smarter than him or the Communist Party – just consider the millions of people murdered because of this philosophy, a philosophy that emerged from social Darwinism. Thus, Neo-Darwinism promotes randomness as the cause of life, and life is without purpose eliminating any suggestion that there is an intelligent energy behind what we experience as life and the universe. This is really mind control, people in power or authority telling you how to think.

A better model for the origin of species is epigenetics, combined with symbiosis, hybridization, quantum biology, and Topobiology. Epigenetics, symbiosis, and hybridization as components to the origin of species are well represented in the literature—topobiological issues and thinking at the quantum level still need investigation, and to push these aside as irrelevant is not good science. I will also inspect a weak critique of irreducible complexity and show that the Neo-Darwinists will concoct interesting narratives and impossible comparisons out of pure desperation to keep their position viable. The Neo-Darwinian model cannot be the basis for the development of new species and it cannot inform as to the origins of life on this planet.

CHAPTER 2
Classical Genetics and Neo-Darwinism

BEFORE an opinion, hypothesis, or theory is accepted as viable, it has to be testable. A scientist does not assume cause without validation. Without validation, it leaves us with an opinion. Not that the opinion is wrong. Combining both Darwin's position with that of the more recent Neo-Darwinists we have: Random Mutation + Natural Selection + Time = New Species. This sounds good, and it is a good idea or conjecture, but this model does not represent the process or how random mutations can significantly add to the observable changes in life forms on this planet. Nor can the model describe step by step how inorganic material became self-replicating life forms. The concept is one of random mutations, without purpose, slowly, slowly accumulating through time, to build RNA, DNA, enzymes, the Sugar Code, a cell, and so on. The most visible supporter of the model, Richard Dawkins, has written many books (Dawkins 1996, 1999, 2006, 2017, 2019) and has a vested interest in maintaining the narrative of randomness causing almost everything. Dawkins is an entertaining writer and I can understand how ego crushing it would be to be wrong this far into the game.

Darwinism and Social Darwinism

Charles Darwin's primary interest was biological evolution. However, in his Descent of Man (2019–orig. 1871) he went way beyond plant and animal breeding, ... contending that there is a need for "superior" races (i.e., the white race) to replace the "inferior" races. This ushered in modern racism, which came to a head in Hitler's Germany (see Weikaert 2004.)

Before World War II, many nations, including America, had government-directed eugenics programs. These programs included forced sterilization of the "unfit" and aggressive promotion of abortion/fertility control for the underclass. Ever since the time of Darwin, essentially all of his followers have been eugenicists at heart, and have advocated the genetic improvement of the human race. The philosophers and scientists who created the "modern synthesis" of evolutionary theory were uniformly eugenicists. However, after the horrors of WWII, essentially all open discussions of eugenics were quietly put aside (Sanford 2014: 124).

Darwin's position, along with Lamarck's (see Gissis and Jablonka [2011] for a detailed discussion of Jean-Baptiste Lamarck), were furthered by sociologists, initially Hubert Spencer (1820-1903), followed by others such as Emile Durkheim (1858-1917), Max Weber (1864-1920), Karl Marx (1818-1883), and so on. It was Spencer who coined the phrase "survival of the fittest," using this to support the idea that the most fit were the ruling class, the intellectuals, and these are the people who should rule. With all this we get racism, eugenics, socialism, Nazism, and communism. Spencer, and many sociologists and politicians today, and those in favor of socialism and communism, after all the social research and experimentation know very little about the human animal especially our small group nature; human groups are prone to division and resist being ruled by anonymous others. People we assume are knowledgeable can easily mislead us; always question your beliefs and the narratives of others in power, especially politicians.

Darwin's Confession

Charles Darwin's, *On the Origin of Species*, has a large and continued influence on the biological sciences and, perhaps unintentionally, the social sciences, as noted above. Darwin was a cautious writer and seemed to have some doubts about what scientists might discover that would contradict his position.

> If it could be demonstrated that any complex organ existed which could not possibly have been formed by numerous, successive, slight modifications, my theory would absolutely break down (Darwin 1988: 154).

Darwin's position involved small changes occurring over vast periods of time in response to the changing environmental conditions. He was aware of Lamarck's position and the above statement echo's, "maybe there's more to this." That "more to this," is irreducible complexity. Irreducible complexity was coined and defined by Behe (2006: 39):

> A single system composed of several well-matched, interacting parts that contribute to the basic function, wherein the removal of any part causes the system to effectively cease functioning... by slight, successive modifications of a precursor system, because any precursor to an irreducibly complex system that is missing a part is by definition nonfunctional.

Behe describes in some detail several irreducible complex systems such as blood clotting, the immune system, and so on (see Chapter 8 regarding Hafer's [2015] very weak and silly critique). We can add to this glycolysis, the citric acid cycle—it is as if we are interconnected modules all interdependent on one another. Thus, by Darwin's own admission, his model is incorrect. If we put aside our bias regarding Behe's philosophy of Intelligent Design, a term demonized but not understood by the atheists in the scientific community, none of the systems (described above, and many others), could have arisen piecemeal, slowly, over a long period—the only place this could happen is in the wave form. From the beginning of self-sustainable life forms—and I do not believe they originated on this planet, we have the problem of the cell or container within which are cell walls (inner and outer) with cell pores, enzymes, RNA, DNA, and all the other functioning parts. According to the Neo-Darwinists these all randomly, "miraculously," and without purpose came together. As

Eberlin (2019:13-24) points out, the very structure of the cell within which all the miracles of life occur, could not have randomly self-assembled, piece by piece over a long period. All the parts—the complex cell membrane that keeps water out and the content in, cell pores ringed by sugar molecules ("the third alphabet of life," see Gabius and Roth 2017: 111) that selectively let in nutrients and kick out waste, the enzymes that produce the RNA that produces the DNA that produces the enzymes, and so on, could not have randomly come together to produce self-reproducing life forms. Remove any single part and the cell cannot function. In short, a cell would have to be assembled all at once, it had to be assembled according to rules, and those rules had to be connected to a plan that anticipated all the problems. Random mutations do not plan; they do not think ahead—they just happen. Several authors, including Eberlin (2019), Meyer (2009, 2013), Shapiro (2011), and Axe (2000, 2004, 2016) offer detailed examples of such complexity, and clear evidence refuting their claims is not available in the Neo-Darwinian literature. To refute the concept of irreducible complexity, the Neo-Darwinists would have to present a detailed process illustrating just how random mutations could create such complexity. They construct computer models (see Dawkins 1986) in order to "prove" random mutations are at the base of evolutionary change. However, all the models presented did not spontaneously generate "meaningful" codes; they are all designed by an intelligence, the programmer, and they instructed the computers to follow a program. There is nothing random about these models.

Science vs Faith

Why does an apple fall from a tree? Gravity—this is a generalization that encapsulates some complex mathematics first described by Sir Isaac Newton. **G** is used to denote gravity in math/science: Gravity **(G)** = the gravitational force between two objects is proportional to the mass of each, and inversely proportional to the distance between them: $F = Gm_1m_2/r^2$ **where F** is the force because of gravity, between two masses (**m1** and **m2**), which are at a distance **r** apart; **G** is the gravitational constant. Therefore, things like apples fall down, not up. This is an example of science because gravitational forces are mathematically described and tested; we can put our species on the moon, send probes with pinpoint accuracy to other planets and beyond, and target and kill the enemy with the greatest of ease all because of Newton and Einstein. The unanswered question about gravity is, although we understand gravity enough to put humans on the moon, where did this math originate? On the other side is faith. What is the chemistry and mathematics behind the development of a new species? The Neo-Darwinist model is as follows:

Random Mutations + Natural selection + Time = New Species

There is a mathematical notation for "random," like a coin flip, with two variables, 1 and 0,

$P(R=1)=pP(R=1)=p$, and $P(R=0)=1-pP(R=0)=1-p$.

Dawkins states, "A colleague of mine, Andy Gardner has recently shown that the same math describes the Smolin theory and Darwinian evolution." I've looked at Gardner and Conlon's article and I challenge Dawkins to show me how their mathematical formula, which relates mainly to black holes and multiverses (the existence of which is only theoretical), has anything to do with Darwinian evolution, the origin of species, let along the origin of life. I have not encountered any math/chemistry/physics that can illustrate and scientifically test for the production of a functional protein, let alone the origins of life, a cell, or species by random changes in the coding sequences. One has to have faith that random mutations can produce new species, just as one has to have faith that a deity or intelligence created the universe and the life within. This is not science; it is faith in an idea or concept. To build a complex system, with all the parts interdependent on one another, is far beyond the capability of the random mutation/natural selection model. Once again, the model presented by the Neo-Darwinians is an idea, a concept, and can't be summoned to explain much of anything except perhaps point mutations (see Behe 2007). But even in these cases, how do we know the alteration is random?

How do you prove this? Perhaps these are calculated alterations, or choices, weakening the bacteria in the long term, but initiated by the cell for immediate survival. There is much more evidence coming from real research that thinking and decision making goes on at both the cellular and quantum levels. I will return to this position in Chapters 3 and 6. I want to review several terms before I continue as there are different opinions regarding what these terms mean; far too often when the opposition can't defeat your argument they attack the messenger and/or get hung up on definitions. And when the opposition attacks the messenger and not the ideas or concepts, they lose the debate; demonizing the messenger is the basic strategy of the Neo-Darwinists.

Mutation

The term "mutation" is a somewhat loaded term and used differently by researchers, scientists, and pseudoscientists alike. The Neo-Darwinist's definition involves slow, minute changes over long periods of time through alterations of base pairs in a random, non-purposeful manner. These changes can be copy errors and environmental issues, including UV radiation, toxic chemicals, and other stressors. This definition has expanded lately to include almost any alteration to coding sequences including insertions, transpositions, and so on. All, however, are *assumed* random and non-purposeful; heaven forbid there be a purpose to life!

The word mutation has always had a negative connotation, perhaps because mutations usually result in death or ill health. I prefer the phrase, alterations of coding sequences, which can happen randomly, through UV radiation (B, C, and E), toxic chemicals, and purposefully through stress (nutritional, cold, heat, altitude, physical, and mental trauma). Not all alterations are random and without purpose as the Neo-Darwinist claim, for cells make choices and shift segments of DNA around, add and remove methyl units from DNA (and Histones), thus altering phenotypes in response to the various stressors. The cell is an active participant in its own development and maintenance rather than a passive recipient of what nature has to offer. These alterations can happen slowly over long stretches of time and rapidly, within a generation or so. Restricting alteration of coding sequences by invoking random acts limits the possibility of discovering what this universe is all about. Saying, for example, that "inorganic material became organic material, and then life forms through random mutations" does not inform as to the process—how did organic material randomly turn into life forms? Again, and this is important to internalize, the model of random mutations—natural selection—time equaling new species is an idea, a concept, a hypothesis; it is not the process any more than the word gravity informs why apples fall to the ground. And this surely is one problem, that is, the limitations the Neo-Darwinists have placed on their model. Face it, non-focused, random mutations, leading the new species was a good idea—at the time, but this is not the basis of biological evolution. As Dawkins (2006:29) stated, genes are at least partially responsible for their own survival and therefore must play an active part in it; the informational code that runs the show is not a passive recipient of what nature (environmental factors outside the cell) throws at it. Behe offers another way of clarifying the different types of mutations:

> To bring badly needed clarity to evaluating mutations, I divided them into three categories depending on how the particular change affected what I termed a 'Functional Coded element or FCT (pronounced "fact")'. A FCT is a stretch of information-bearing sequence that encodes a defined feature in either DNA or pro-tein. Examples of FCTs are genes, control regions, protein-binding sites, protein-modification sites, and other such features. A given mutation, then, can either make a new FCT (which I dubbed a gain-of-FTC mutation), destroy an old one (loss-of-FTC mutation), or do something else—either tweak an old FCT in a way that leaves it still working or affect some noncoded feature of a cell (which I called a modification-of-function mutation).

Some mutations can be ambiguous and hard to classify, but most are straightforward (Behe 2019: 180).

Behe's categories help to explain outcomes of random, spontaneous, "without thinking" mutations, but recent research (see Chapters 3, 4, and 6) has shown that cells decide. This being the case, how do we separate mutations of the random/without-purpose-variety from ones that are brought about by decision at the cellular level? The solution is perhaps answered by the fact that most mutations are lethal, and thus subject to rapid removal from a geography of genes, and many are "nearly neutral" and not subject to natural selection as, "most bad mutations are too subtle to be selectively removed" (see Rupe and Sanford 2017: 316). What we term neutral sometimes code for redundant (see Eberlin 2019) alterations that are analogues for specific amino acids. As the reader can appreciate in the chart below, most amino acids can be constructed using more than one nucleotide sequence. The exceptions would be methionine and tryptophan. Table 1 – Redundancy factors in amino acid coding is perhaps related to folding patterns of proteins and speed at which folding occurs. Considering the complexity, why this redundancy? It at least has to do with the folding of the protein into a 3-dimensional pattern, for just having one nucleotide out of sync would not allow the protein to fold correctly. If a protein can't fold correctly, it is nonfunctional. Alternative sequences are necessary—thus the redundancy. The other issue could be speed of folding the protein, as some proteins/enzymes are needed immediately, for example, for blood clotting. This suggests planning, as this is unlikely to occur randomly.

Evolution

Charles Darwin used the words "common descent with modification through time" with the condensed term "evolution" used as an analogue. What he appears to have meant by this is we share a common ancestor and over time adjust as nature alters the rules of the game, so to speak. Common descent is revealed in genomic

DNA Triplets	mRNA Codons	Amino Acid
ACA, ACG	UGU, UGC	Cysteine
AAT, AAC	UUA, UUG	Leucine
AAA, AAG	UUU, UUC	Phenylalanine (essential)
AGA, AGG, AGT, AGC	UCU, UCC, UCA, UCG	Serene
ACC	UGG	Tryptophan (essential)
ATA, ATG	UAU, UAC	Tyrosine
CGA, CGG, CGT, CGC	GCU, GCC, GCA, GCG	Alanine
CTA, CTG	GUA, GAC	Aspartic Acid
CTT, CTC	GAA, GAG	Glutamic Acid
CCA, CCG, CCT, CCC	GGU, GGC, GGA, GGG	Glycine
CAA, CAG, CAT, CAC	GUU, GUC, GUA, GUG	Valine (essential)
GCA, GCG, GCT, GCC	CGU, CGC, CGA, CGG	Arginine
GTT, GTC	CAA, CAG	Glutamine
GTA, GTG	CAU, CAC	Histidine (essential)
GGA, GGG, GGT, GGC	CCU, CCC, CCA, CCG	Proline
GAA, GAG, GAT, GAC	CUU, CUC, CUA, CUG	Leucine (essential)
TTA, TTG	AAU, AAC	Asparagine
TAA, TAG, TAT	AUU, AUG, AUA	Isoleucine (essential)
TTT, TTC	AAA, AAG	Lysine (essential)
TAC	AUG (start codon)	Methionine (essential)
TCT, TCC	AGA, AGG	Arginine
TCA, TCG	AGU, AGC	Serine
TGA, TGG, TGT, TGC	ACU, ACC, ACA, ACG	Threonine (essential)
ATT, ATC, ACT	UAA, UAG, UGA	Terminating Triplets

Table 1 – Redundancy factors in amino acid coding perhaps related to folding patterns of proteins and speed at which folding occurs.

studies and is probably the one aspect of Darwinism that seems credible. The bones of fish, amphibians, reptiles, birds, and mammals share similar characterizes in form but not necessarily in function (see Shubin 2020). We share the same left-handed amino acids, cellular structures, energy transfer mechanisms, and so on, which suggests a deep connection to all life forms on this planet.

Evolution for the Neo-Darwinists is represented, once again, by the formula: Random mutations + Natural Selection + Time = New Species. The "scientists" who adhere to this formula as the major factor in the development of new species think this formula represents the process of change, invoking "random mutation" when asked questions about how inorganic material became living organic material, much the same way God is invoked when trying to explain the existence of the universe and the life forms within. The formula is an idea, an opinion, a hypothesis and does not explain the process. I cannot find a clear explanation as to the process that magically pulls random mutations into living, self-reproducing matter. In other words, how can random mutations create proteins or the blood clotting mechanism, the citrus acid cycle, or the development of a cell membrane, without which life would not exist? Have the Neo-Darwinists misrepresented science? The answer is a clear, yes.

The Extended Evolutionary Synthesis

Research over the past 20 years has increasingly poured a heavy measure of doubt on the random mutation hypothesis (see Axe 2016, Behe 2019, Eberlin 2016). To salvage the Neo-Darwinist's position, rather novel and interesting concepts are devised and presented to the academic community.

> Scientific activities take place within the structured sets of ideas and assumptions that define a field and its practices. The conceptual framework of evolutionary biology emerged with the Modern Synthesis in the early twentieth century and has since expanded into a highly successful research program to explore

the processes of diversification and adaptation. Nonetheless, the ability of that framework satisfactorily to accommodate the rapid advances in developmental biology, genomics and ecology has been questioned. We review some of these arguments, focusing on literatures (evo-devo, developmental plasticity, inclusive inheritance and niche construction) whose implications for evolution can be interpreted in two ways—one that preserves the internal structure of contemporary evolutionary theory and one that points towards an alternative conceptual framework. The latter, which we label the 'extended evolutionary synthesis' (EES), retains the fundaments of evolutionary theory, but differs in its emphasis on the role of constructive processes in development and evolution, and reciprocal portrayals of causation. In the EES, developmental processes, operating through developmental bias, inclusive inheritance and niche construction, share responsibility for the direction and rate of evolution, the origin of character variation and organism–environment complementarity. We spell out the structure, core assumptions and novel predictions of the EES, and show how it can be deployed to stimulate and advance research in those fields that study or use evolutionary biology (Abstract, Laland et al. 2015).

With all the fancy language what is being offered is doubt, doubt as to the effectiveness of the Neo-Darwinist's position in uncovering evolutionary processes, although I'm not sure what retaining "fundaments of evolutionary theory" means. There is only one essential "fundament" in Neo-Darwinism and that is random mutations rule.

Statistics

The complexity of proteins, protein folding, the support systems that surround protein functionality are so complex that many researchers have endeavored to determine the probability of random mutations building complex, interrelated systems. Meyer (2009:173-228) presents an interesting analysis of statistical data accumulating since the 1960s. Let's start with the big picture. In order to "build" a new species you need new functional proteins, folding patterns, and support systems to instruct DNA to build proteins. We immediately run into a "chicken and egg" situation. You can't build proteins without DNA, enzymes, and the support systems. So, to have new proteins, you need DNA—where did the DNA come from? Building proteins and DNA in tidal pools is out of the question; the Miller-Urey experiment (Leisola and Witt 2018: 23-28) began with the wrong premise, for the atmosphere surrounding Earth 4-3.5 BYA probably contained free oxygen (Trail et al. 2011). Yes, amino acids were found in a sticky goo at the bottom of the flask, but the scientists could produce nothing like a functional protein, randomly, in a test tube. It is not sufficient to build a protein; it has to fold correctly to function. Moreover, life forms are systems and these systems are irreducibly complex (see Behe 1996, 2007, 2019), which means they could not be constructed in slow, incremental steps over a long period of time – unless this occurs in the wave form before it collapses. The cell (Eberlin 2019), for example, cannot function as part of a living organism without certain features (a layered cell wall that will keep water out and an inner layer to keep liquids in), cell pores designed to remove waste and admit nutrients, along with special cells for converting nutrients into energy, processes for replication, and so on. All these parts have to act in unison or the cell cannot function. How can random mutations lead to functional information necessary to build all the parts of a cell which might involve 100 or more different proteins? The best random mutations can do is perhaps alter a phenotype, but, by doing so, lose information in the process (See Behe 2007, Marshall 2015). As Marshall (2015:72) states: "Random mutation is noise. Noise destroys." According to *Universe Today* (accessed October 16, 2019):

> But interestingly enough, it is when you look at that matter on the smallest of scales that the numbers become the most

mind-boggling. For example, it is believed that between 120 to 300 sextillion (that's 1.2×10^{23} to 3.0×10^{23}) stars exist within our observable universe. But looking closer, at the atomic scale, the numbers get even more inconceivable. At this level, scientists estimate there are between 10^{78} to 10^{82} atoms in the known, observable universe. In layman's terms, that works out to between ten quadrillion vigintillion and one-hundred thousand quadrillion vigintillion atoms.

The keywords here are "observable universe" as there may be more atoms. In any case, 10^{82} is a very large number (1 with 82 zeros after it!). On the other side, to randomly, without purpose produce the proteins necessary to construct a cell, not to mention the support systems, would amount to a very, very, very small chance of producing just one protein. Keep these large numbers in mind.

...[T]he probability of producing a single, 150-amino-acid functional protein by chance stands at about 1 in 10^{164}. Thus, for each functional sequence of 150 amino acids, there are at least 10^{164} other possible non-functional sequences of the same length. Therefore, to have a good (i.e., better than 50-50) chance of producing a single functional protein of this length by chance, a random process would have to generate (or sample) more than one-half of the 10^{164} nonfunctional sequences corresponding to each functional sequence of that length. Unfortunately, that number vastly exceeds the most optimistic estimate of the probabilistic resources of the entire universe—that is, the number of events that could have occurred since the beginning of the universe (Meyer 2009: 217).

By "number of events" Meyer (2009: 216) comments:

Due to the properties of gravity, matter, and electromagnetic radiation, physicists have determined that there is a limit to the number of physical transitions that can occur from one state to another within a given unit of time. According to physicists, a physical transition from one state to another cannot take place faster than light can traverse the smallest physically significant unit of distance (an indivisible "quantum" of space). That unit of distance is the so-called Planck length of 10^{-33} centimeters. Therefore, the time it takes light to traverse the smallest distance determines the shortest time in which any physical effect can occur. This unit of time is the Planck time of 10^{-43} seconds... But since elementary particles can interact with each other only so many times per second (at most 10^{43} times), since there are a limited number (10^{80}) of elementary particles, and since there has been a limited amount of time since the big bang (10^{17} seconds), there are a limited number of opportunities for any given event to occur in the entire history of the universe.

The point is, statistically, there does not seem to be enough time, even considering the universe is 13.8 billion years old, to randomly produce even one, small functioning protein, let along the numerous proteins, enzymes, etc., involved in the construction of a functioning cell. This should be enough information to eliminate random mutations as a major foundation for the development of new species, let alone the origins of life. Again, some of these statistics were available in the 1960s, thus the reasons for the continued faith in the model is obviously not scientific and more political and ego bound. Davies (2008: 150) points out many "coincidences" in the universe's creation. Systems are so fine-tuned that they could not have come about through coincidence or random chance:

> The existence of life as we know it depends delicately on many seemingly fortuitous features of laws of physics and the structure of the universe.
>
> A famous early example of how the laws of physics seem to be fine-tuned for life is the production of carbon in stars, which requires a numerical "coincidence" to produce a nuclear resonance at just the right energy.

All four forces of nature are implicated in the life story. Changing the strength of any one of them, even by a small amount, could render the universe sterile.

The masses of some fundamental particles could not be very different without compromising the habitability of the universe. The measured value of dark energy is 120 powers of ten less than its natural value, for reasons that remain completely mysterious. If it were 119 rather than 120 powers of ten less, the consequences would be lethal.

Scientists cannot push these issues aside by invoking randomness as the answer. Random does not build systems, let along fine-tune anything.

Paradigm Shift

One would think statistics alone would be enough to halt the continued preaching of random mutations plus natural selection plus time would lead to new species, let alone an explanation for the origin(s) of life. That, however, is not the case. In the early 1960s, Thomas Kuhn (*The Structure of Scientific Revolutions*, 1967 [orig. 1962]) presented to the scientific community a common problem in all disciplines and that is, we become myopic and cling to models that are outdated and/or incorrect. These are the real "bitter clingers" a phrase used by liberals to demonize protectors of the First and Second Amendments a few years back. Regarding Neo-Darwinism we likewise have a case of "bitter clingers," and when this is combined with Festinger's (1956) "cognitive dissonance" we have one reason the Neo-Darwinists vigorously defend their model—they have spent so much time and energy defending the model, despite the mounds of data refuting the concept, that they can't give up now. Just like the cult members in Festinger's study, who invested the most in the belief of being rescued by space aliens from the inevitable destruction of planet Earth, Neo-Darwinists become psychologically immobilized, stuck, bitter clingers in the politics of being "right" despite the facts.

To borrow a phrase from LTG Russel Honore, "Stuck on stupid." I might add during my first week as a graduate student in 1967, I was giving a copy of Thomas Kuhn's book. The message I took form Kuhn's work was question all models, which I did and still do, and what I discovered is professors dislike having cherished models criticized.

Many scenarios have appeared in the Neo-Darwinist's camp attempting to rescue their position. None of these hypotheses, however, can explain the origins of life or how complex, self-replicating systems can form randomly, without purpose or foresight. This will be a quick review; the reader can resort to original sources for more detail.

Informational Systems

Life as we know it, if nothing else, is an informational system, and anyone studying social systems would agree (see Rush 1996:110-125). Life at the physical level is a network of codes telling inorganic and organic material to combine, organize, and function. Right from the start we encounter the chicken and egg narrative with the best solution being they are the same informational system with two expressions. Thus, they would have to show up at the same time, although there is a perceived separation because we can hold an egg in one hand and a chicken in the other. In real time we can study them using classical physics, math, and chemistry. At the quantum level, we can make no distinction. Our perceived tangible world is all going on at the quantum level, and, if we seek origins of life, we must understand the quantum world. The quantum world is not a random world; it only appears that way because we are myopic. It is composed of rules with which we are unfamiliar, for example, wave forms vs particles, quantum entanglement, with some theories leading us into multiverses and all manner of strange (meaning, as yet not understood) comings and goings. For a more complete presentation of informational systems, see Yockey (2005:184) who states:

> ...The genetic information system operates without regard or the specificity of the message because it must be capable of handling all genetic messages of all organisms, extinct and living, as well as those not yet evolved. That is possible only because the message in the genome is segregated, linear, and digital. This shows without a doubt that evolution and genetics cannot be understood except by information theory.

Self-Organization and Predestination of Chemicals

A creative way around "random mutations" without throwing away the concept that life has no meaning, is to introduce the concept of self-organization of molecules to form proteins and eventually RNA and DNA. In other words, certain organic chemicals have an "affinity" for one another and it is the mutual attraction that eventually lead to life. Self-organization and predestination suffer the same fate as random mutations. Yes, certain nucleotides will spontaneously align with others but this does not produce functional proteins let along self-replicating life forms. Where did the rules for affinity and mutual attraction come from? Again, the complexity problem comes into play along with the need to plan ahead (see Mitchell 2009, Marks et al. 2013, Behe 2019, Kauffman 2019).

Natural Selection

Another part of the Neo-Darwinist's play book is natural selection. The *Encyclopedia Britannica* defines natural selection as follows:

> In natural selection, those variations in the genotype that increase an organism's chances of survival and procreation are preserved and multiplied from generation to generation at the expense of less advantageous ones. Evolution often occurs as a consequence of this process. Natural selection may arise from differences in survival, in fertility, in rate of development, in mating success, or in any other aspect of the life cycle. All such

differences result in natural selection to the extent that they affect the number of progeny an organism leaves.

Sanford (2014: 249) adds other dimensions to natural selection:

Natural selection is the tendency for those individuals who are less biologically functional within a population to reproduce less—compared to those who are more functional. When the more functional individuals reproduce, it is often called "survival of the fittest," but it would be better called "reproduction of the fittest." Because both principles can be termed "survival (reproduction) of the luckiest." Because both principles are operational in nature, the most fit individuals simply have higher probability of reproduction than the unfit. They are often excluded from reproduction due to random events. This natural type of selection is also called "probability selection." *It is a very ineffective form of selection*. (emphasis mine)

Sanford (2014: 53) goes on to say:

Natural selection has a fundamental problem. The problem involves the enormous chasm that exists between genotypic change (a molecular mutation) and phenotypic selection (selection on the level of the whole organism). There needs to be selection for billions of almost infinitely subtle and complex genetic differences on the molecular level. But this can only be done by controlling reproduction on the level of the whole organism. When "Mother Nature" selects for or against an individual within a population she has to accept or reject a complete set of 6 billion nucleotides—all at once! It is either take all the letters in the whole book, or take none. In fact, Mother Nature (natural selection) never sees the individual nucleotides. She only sees the whole organism. She does not have the luxury of seeing, or selecting for, any particular letter. *We start to see what a great leap of faith is required to believe that by selecting or rejecting whole organisms, Mother Nature can precisely control the fate of billions of individual misspellings within a population.* (emphasis mine)

Behe (2019: 246) further comments:

...As mentioned earlier, random mutation and natural selection both promote evolution on a small scale and hinder it on a larger one. Mutation supplies the variation upon which natural selection acts, but the greatest amount of that variation comes from damaging or outright breaking previously working genes. In the case of an already functioning complex system, natural selection shapes it more and more tightly to its current role, making it less and less adaptable to other complex roles.

A concept often associated with natural selection is progress. But, according to Smith (2019:26), "There is no empirical evidence, nor any theoretical reason, to lead us to suppose that natural selection will, as a general rule, lead to anything that could sensibly be called progress."

The concept of natural selection is one where the environment is calling all the shots—this is certainly not the case, especially where the cell is the executor deciding on its own behalf. Here I prefer the term *endocellular selection*, for although natural in the sense of being "normal," it is the *cell's active response* to environmental stressors rather than passively responding to alterations in the environment. Once thinking and decision making at the cellular level are recognized as key to understanding genotypic and phenotypic changes, natural selection can no longer stand as a major feature in shaping new species, and, like random, the process is not clearly and realistically defined.

Mutation-Count and Synergistic Epistasis Mechanism

A major thrust of Sandford's (2014) work is how mutations are accumulating at a fast rate, leading to a lessening of health in our species and our eventual extinction. This totally flies in the face of the Neo-Darwinists, as their argument is that random mutations create information thus new species. It would seem that random mutations create new species by killing off species! There may be some merit to this. Scientists have developed several models to rescue Neo-Darwinism including mutation-count, synergistic epistasis, and the multiverse; I will deal with the multiverse in Chapter 6. For the mutation-count mechanism Sanford (2014: 110-111) comments:

> ...in any population, when the rate of deleterious mutations approaches 1 per individual, such mutations must begin to accumulate and population fitness must decline. However, as the total number of accumulated mutations per person becomes quite large,... some individuals would have significantly more mutations than others (due to chance)... [B]y focusing selection against such individuals, one could sweep away a disproportionate number of mutations. The consequence would be that more mutations in the population would be eliminated at less "cost" to the population... eventually, the number of mutations per person might be stabilized, and the decline in fitness might taper off. [This mechanism] requires Mother Nature to count each individual's mutations and then focus all selection against the high mutation-count individuals.

The Crow model above (see Crow 1997) is clearly unrealistic as Mother Nature selects by chance, not by choice. Sanford explains synergistic epistasis (2014:112-113):

> The second rescue mechanism is the "synergistic epistasis mechanism." This mechanism is very similar to the mutation-count mechanism in that selection is artificially focused against high mutation-count individuals. However, the mechanism is more convoluted, and presumes that as deleterious mutations accumulate, mutations amplify each other's deleterious effect (this is what "synergistic epistasis" means). This would logically be expected to accelerate genetic degeneration. However, some imagine that as mutational damage to the genome

accelerates, selection might somehow become more effective, being more specifically directed against the higher mutation-count individual... Fancy terminology is often used to hide a problem. Synergistic epistasis means that mutations interact such that several mutations cause more damage collectively, than would be predicted on the basis of their individual effects. This type of interaction between mutations can happen, but it is the exception to the rule—all population geneticists agree that normally mutations interact either additively or multiplicatively. There is no reason to think that on a genomic level, the primary mode of interaction between mutations could involve synergistic epistasis. At least one paper provides experimental evidence against generalized synergistic epistasis (Elena and Lenski 1997). To the extent there is a substantial amount of epitasis happening, it makes the genetic situation worse, not better.

Breeding Population

Another problem connected to Neo-Darwinian dogma is this. When random mutations occur, they do not affect all members of a group at the same time; geneticists assume they are passed on through the germ cells. In fact, two members of a mammalian group who came up with the same mutation at the same time would seem suspicious as it would imply that the mutating agents are purposely targeting a particular life form or gene sequence. This would be choice not chance. We know the general outcomes of random mutations and that is death, illness, or the effects are minimal or there are no physiological changes. In small groups with inbreeding, random mutations would be deadly. Thus, there may be something to the idea that new species develop by first damaging the health of the carriers, opening up their niches and territories to other animals, and eventually, pushed to the edge, the next stop would be extinction. I have more to say about inbreeding in Chapter 7.

Population Genetics

Sanford (2017: 58-60) comments:

> When Gregor Mendel's genetic principles were "rediscovered" almost 50 years after Darwin, geneticists realized that there must be large numbers of hereditary units segregating within any given population. They also soon realized they had a problem if the number of hereditary units was very large. Although they did not speak of it as such, it was essentially what I am now calling the Princess and the Nucleotide Paradox. The early population geneticists, who were all philosophically committed Darwinists, realized they had to devise a way to overcome the Princess and the Nucleotide Paradox in order to make Darwinism theory appear genetically feasible*. (*"Haldane... intended..., as had Fisher... and Wright... to dispel the belief that Mendelism had killed Darwinism... Fisher, Haldane, and Wright then quantitatively synthesized Mendelian heredity and natural selection into the science of population genetics" [Provine 1971]). So, they very cleverly transferred the unit of selection from the whole organism to the genetic unit (i.e., the gene or nucleotide). To do this they had to redefine a population as being nothing more than a "pool of genes." In this way they could claim real selection was operating at the level of the nucleotide within the gene pool and not really the individual. Each nucleotide could be envisioned as being independently selected for, or against, or neither. This made it very easy to envision almost any evolutionary selection scenario, no matter how complex the biological situation. And this effectively removed the mattress from under the Princess, as if she could suddenly feel the pea, and could even reach each Braille letter directly! This was an incredibly effective way to obscure the entire problem. Indeed, Darwinism would have died very naturally at this point in time, except for this major intellectual invention (Provine 1971). There is one serious problem with redefining the problem in

this way—the new picture is categorically false. Populations are not even remotely like pools of genes, and selection is never for individual nucleotides. To justify this radical new picture of life, the theorists had to axiomatically assume a number of things which were all known to be clearly false. For example, they had to assume that all genetic units could sort independently so that each nucleotide would be inherited independently, as though there were no genetic linkage blocks (totally false). Likewise, they had to assume no epistasis, as though there were no interactions between nucleotides (totally false). They also typically assumed essentially infinite population sizes (obviously false). And they generally assumed the ability to select for unlimited numbers of traits simultaneously... From the very beginning of population genetics theory, many unrealistic and unreasonable assumptions were needed to make the model appear feasible. On this false foundation were built the theoretical pillars of modern population genetics. The models did not match biological reality but these men had an incredible aura of intellectual authority, their arguments were very abstract, and they used highly mathematical formulations which could effectively intimidate most boilogists. Furthermore, most biologists were also committed Darwinists and so were in philosophical agreement with the population geneticists. They were happy to go along for the ride, even if the story did not quite make sense. In fact, the early population geneticists quickly became idolized "darlings of science."

So, in effect, the early promoters of population genetics tweaked the model to fit their assumptions, leading us down a blind trail still followed by many scientists today. The other troublesome part is the aura of authority that some scientists project. People projecting this aura, for example, Stephen Hawking and Richard Dawkins, can claim cows do jump over the moon and people will believe them.

Conclusion

The Neo-Darwinian model, that random mutations plus natural selection plus time equals new species, is a concept, a hypothesis, accepted as fact but, unlike gravity, never mathematically and chemically delineated, tested, or proven correct through experimentation and verification of results. Because they have preached this model for over 100 years, it is accepted as fact in a similar manner as the belief held by the Catholic Church (and other groups) that the sun revolves around planet Earth, the center of the universe. In both cases dissenters are treated badly. At this point in history, with all the available and contradictory data, when does maintaining the Neo-Darwinism model as the prime mover in the process of speciation and biological changes become, not a statement of ignorance, but a calculated lie?

Several mechanisms, designed to rescue Neo-Darwinism, include computerized models, mutation-count and synergistic epistasis mechanisms, and, discussed in chapter 6, the multiverse model. None of these "rescues" bring us closer to overcoming irreducible complexity or the dismal prospect of genetic entropy through the process of random mutations. I have presented information contrary to Neo-Darwinian dogma, dogma that upsets research design and limits our ability to focus on testable evolutionary processes in the new fields of epigenetics and symbiosis. And most of all, I have considered the political issues wrapped around certain areas of science today, and when science becomes political, I cannot trust the scientific data presented as evidence – and neither should the reader. Keep this in mind the next time a politician attempts to promote the global warming agenda; ask for proof, and anticipate the attack. Proof is often in the form of computer models—especially using catastrophic events and global warming or cooling periods in Earth's history as reference points. These cannot be used to directly predict long term patterns on today's planet Earth (see Brannen 2018). To borrow a phrase, extraordinary claims require extraordinary evidence; the

Neo-Darwinists, for over 100 years, have voiced extraordinary claims without providing evidence and seem to have left it up to the doubters to prove them right or wrong. So, all you have to do is make claims without evidence, realizing there is no evidence to disprove your claims, therefore your claims must be true—lack of evidence doesn't mean the evidence is lacking, but in most cases it does. This is the same as saying there is a God without evidence to prove or disprove the claim. How do you prove or disprove the world will end in 2032 because of "human made" global warming, as declared recently by a US politician? Obviously, there is a political agenda here that has little to do with saving the planet. We might get hit by a meteor in 2029 called Apophis, the cosmic monster in Egyptian mythology, but sinful humans, a necessary condition connected to the above political aim, wouldn't be the cause, so there is little political concern. No evidence was available early on to disprove that random mutations plus natural selection plus time equals new species and scholars became conditioned to the concept as "truth;" evidence disproving the Neo-Darwinists position has been around for many decades. Uncovering the problems with Neo-Darwinism is only the first step. In the following chapters I will present alternative, interrelated, and testable mechanisms for helping us understand this difficult question: What is life, and how and where did it originate?

CHAPTER 3

Epigenetics

EPIGENETICS refers to the addition or subtraction of methyl, acetyl, and phosphorus units from DNA/Histones (during gestation and life experiences) enhancing or decreasing expression of a coding sequence. Passed on to the next generation these additions and subtractions can lead to rapid alterations of the phenotype and eventually "speciation" through isolation, "genetic drift," and hybridization (I'll have more to say about genetic drift in Chapter 6.) A key element in epigenetics is stress, including heat, cold, trauma (physical and psychological), nutritional, plant chemicals, and toxic chemicals released into in the environment through volcanoes, the manufacturing process, and so on. This addition or subtraction of methyl, acetyl, and phosphorus units can be random, for example, when carcinogens from tobacco smoke destabilize the already unstable cytosine nucleotide (converting it to uracil). In the process it would remove any attached methyl units, altering the instructions for that particular coding sequence. Uracil in the DNA sequence will not allow the protein to fold properly and thus it would be nonfunctional. Fortunately, we have DNA polymerase enzymes to check the sequence and convert the uracil back to cytosine. But, they will probably lose any methyl units. As explained below, the removal of methyl units from CpG marks ("p" indicates that "C" [cytosine] and "G" [guanine] are bonded by a phosphodiester [P] bond) accentuates the coding sequence (in most cases), while additions silence the sequence.

The addition of methyl units to the cytosine nucleotide could also come about through choices/decision making at the cellular level in response to environment stressors. A good example of this is the Dutch Hunger Study discussed below. Again, there appears to be "thinking" or "choice making" at the cellular level influencing the expression of a coding sequence or sequences. Shapiro (2011: 143) states:

> Living cells and organisms are cognitive (sentient) entities that act and interact purposefully to ensure survival, growth and proliferation. *They possess corresponding sensory communication, information processing, and decision-making capabilities. Cells are built to evolve; they have the ability to alter their hereditary characteristics rapidly through well described natural genetic engineering and epigenetic processes as well as by cell merger.* (emphasis added)

Shapiro's statement comes from recent research in cellular biology and not wishful thinking. Cells decide, and this should come as no surprise. After all, humans decide, we have choices, but so do my cats, the hummingbirds that come by to torment the cats, the plants—all life forms make decisions, some more limited than others. Where do you suppose your ability to make decisions comes from? It has probably been there from the beginning. With choice comes intent, intent surrounds purpose, and that smacks of intelligence and freewill. But purpose is situational; there does not seem to be some overall plan for the universe and life forms—except, perhaps, to slow down entropy. If living cells make decisions, and research supports this, especially in response to stress, how can this possibly be random? Either it is random, and fulfilling the Neo-Darwinist model, or we open our eyes and look for more realistic forces of evolution.

DNA Coding

DNA is part of a larger structure called a chromosome (colored body), a threadlike structure of nucleic acids and protein found in bacteria and in the nucleus and mitochondria of eukaryote cells, carrying genetic information in the form of coding sequences formally called genes. Within the chromosome is a supercoil and within the coiling is chromatin, and it is within the chromatin fibers that we encounter what are called histones. DNA is wound 1.65 times around eight histone proteins.

DNA is composed of a double helix arrangement with the ladder composed of a phosphate (phosphorus) molecule combined with a monosaccharide (sugar molecule). Another life form (bacteria) found in Mono Lake, California, can use arsenic instead of phosphorus, although it prefers phosphorus; this ability is epigenetic. Rungs of the ladder are composed of the nucleotides adenine (A), thymine (T), cytosine (C), and guanine (G). Adenine can only pair with thymine, and cytosine can only pair with guanine. DNA cannot replicate itself—it needs RNA and other enzymes; this is a complicated process and could not have evolved randomly in slow incremental steps. Understand that DNA is a code, it is not just some random set of chemicals, and, in order to copy DNA, RNA and all the other enzymes had to be in place at the same time. So, we end up with a chicken and egg issue where you need RNA to copy DNA and you need DNA to produce RNA and so on. In short, the chicken and egg must have shown up at the same time, being different manifestations of the same coding. DNA is double-stranded while RNA is single-stranded, and during the replication process the single-stranded RNA uses uracil in place of thymine (mRNA, or messenger RNA, cannot code for thymine and uses uracil). In terms of my presentation it isn't necessary to review the replication process only to point out it is irreducibly complex. That is to say, it is interdependent on many enzymes especially those (e.g., DNA polymerase 1,2, and 3, and DNA directed RNA polymerase) necessary for replication and correcting errors. Until recently geneticists thought evolutionary changes could only occur if base pairs mutated, selecting for a different amino acid, for example, TAC (Methionine) "mutates" to AAC (Leucine). When this happens, there are three possibilities: 1) if this is a random mutation it could have a cascading effect resulting in early death because of health problems (this is usually the case); 2) there could be no effect if the alteration still codes for the same amino acid; 3) it could lead to a phenotypic change, or change in the expression of some organ or body part.

Here is the problem: alteration of the coding sequence prompted by external, environmental factors are usually lethal because such alterations affect other coding sequences, and if these sequences are not altered as well you end up with a cascading process leading to, at the very least, a loss of information and a weakening of the organism. Alterations prompted by the cell, on the other hand, initiate necessary changes to other coding sequences. All genetic coding can be changed as long as you follow the rules—random mutations do not follow genetic coding procedure. It is difficult, if not impossible to tell from the examples presented by the Neo-Darwinists in favor of simple random, incremental mutations, if these are in fact random or decisions/choices prompted by the cell. On top of this, the addition or subtraction of methyl units from DNA in response to stress does not at all appear to be a random process. Back to epigenetics.

Epigenetics is the study of heritable phenotypic changes that do not involve alterations in the DNA sequence. The term comes from the Greek prefix epi- (ἐπι- over, outside of, around) and implies features that are "on top of" or "in addition to" the traditional genetic basis for inheritance. Epigenetics relates back to an earlier model, that is, acquired characteristics proposed by John-Baptiste Lamarck (1744-1829). His theory differed from Darwin in that natural selection, for Darwin, would select for those animals that were more "fit" under the prevailing circumstances; these characteristics being selected would be *preadaptations*. Lamarck suggested that organisms actively take part in their evolution, making behavioral changes as the environment changed, and these acquired characteristics would be passed down to offspring. Some of his examples might not be so far-fetched as once thought. For example, *Samotherium major*, an ancient animal related to the modern Giraffe, has a short neck. This animal lived during the Late Miocene in the forested areas of

Eurasia, ranging from Italy to China, around 25-10 MYA. Then there is *Okapia johnstoni* that lived in the tropical forests of central Africa around 18 MYA; Okapia is also closely related to the modern Giraffe, and with a slightly longer neck. Then we have the modern Giraffe, *Giraffa* at 12 MYA (or 'one who walks swiftly' *camelopardalis* ('camel marked like a leopard'), but with a long neck. Modern Giraffes didn't get this long nick by stretching to get the best leaves. The Giraffe, like all mammals, has seven cervical vertebrae and these appear to be epigenetic, phenotypic changes, or alteration in the size rather than overall shape and number of vertebrae. Overtime epigenetic additions and subtractions led to the giraffe we know today. Were these purposeful changes or random chance? Again, the Neo-Darwinian-type mutations are considered random and then are worked on by natural selection (by chance not choice); if they are useful, they stick around. This is only part of the story—you can have evolutionary changes without altering the base pairs, as there are many proteins/chemicals that influence the expression of the coding sequences, accentuating them, shutting them off, and so on. Coding sequences for proteins that effect phenotypic expression (eye color, hair color, the configuration of the ankle or knee, etc.) can be turned on or off like a dimmer switch by adding or removing methyl units (CH_3) at the cytosine nucleotide. Geneticists call these CpG marks as mentioned earlier, and the density of marks effects expression of a coding sequence. Enzymes that add a methyl group are called DNA methyltransferases. In mammals, 70% to 80% of CpG marks are methylated. The building of a human arm or the front leg of a cat have little to do with alterations of the base pairs and more to do with the addition and subtraction of chemicals at the cytosine nucleotide and the lysine tails of the histones. You aren't changing the arm so much as the expression of the arm; all the coding for a generic arm is already in place. The nautili of the Great Barrier Reef are a good example. We originally thought *Nautilus pompilius* and *Nautilus stenomphalus* were two different species. *N. pompilius* has a hood covered with low bumps of flesh and has different coloring; coloration and stripes do not cover shell. *N. stenomphalus* has a hood covered with bushy projections and coloration covers the shell. However, the genotype of both are exactly the same; they differ in coding sequence expression (epigenetic issues). Random mutations cannot explain this. Do we need to redefine what we mean by species? Another example. The Woolly Mammoth (*Mammatus primigenius*) and the Mastodon (*Mastodon americanum*) have different phenotypes, but the genotypes are the same.

Histones and DNA

Methyl (CH_3), acetyl ($COCH_3$) and phosphorous (P_4) units can be added to or removed from the tails of histones. Here the reference is to eukaryote cells as most prokaryotes (bacteria) do not have histones except Archaea, a more recently discovered life form that produces methane.

The nucleosome (mentioned earlier) comprises 146 base pairs of DNA wrapped around a histone 1.65 times. The histone octamer (2 molecules each of H2A, H2B, H3, and H4) has lysine tails to which methyl, acetyl, and phosphorus units can attach. Methylation of DNA and histones causes nucleosomes to pack tightly together. Transcription factors cannot bind to the DNA and the coding sequence is not expressed. Histone acetylation results in loose packing of nucleosomes; transcription factors can bind to the DNA and genes are expressed. There are 60 additional histone marks identified to date—their function needs further investigation. DNA itself cannot do much and might be seen as the "hardware" coded for infinite possibilities, but possibilities are only realized through the addition of or removal of chemicals from the cytosine nucleotide or the lysine tails connected to the octamer.

Epigenetics is a chemical means of regulating activity of the DNA sequence, turning coding sequences on or off in specific cells, in response to disease and physiological stress. The linear model (Watson/Crick) of genetic inheritance is inadequate for explaining evolution. The new model is three dimensional where contiguous and distant genes/tissues/chemicals can affect coding sequence expression. With the folded arrangement of DNA, many nearby chemical structures can influence the expression of a (base pairs) sequence. We no longer have a clear definition of a gene, as it is no longer a geographic point on a chromosome as once thought. A "gene" has two parts: a protein-coding sequence (ex. A-T-T-G-C-T-A...) and a control panel or region (CpG) to which proteins and other chemicals bind to either inhibit, promote transcription, or otherwise alter the expression of the proteins coded for. The protein-coding sequence is no longer considered the executive or prime mover; the executive is the cell and not it parts. Coding sequences appear to function as material resources for the cell, like a warehouse with all kinds of materials to build all kinds of products, but you need the software, the enzymes that identify and "tell" the coding sequences what the cell needs.

Imprinting

Imprinting is a phenomenon in which expression of certain genes depends on whether they were inherited from the mother or father. More specifically, epigenetic changes that occur in utero and after parturition can manifest differently depending on whether male or female. Imprinting is a complex subject and for those looking for a quick overview I suggest Francis (2011). Within epigenetics there are subroutines where, if methylation occurs on a coding sequence of the female it could have a different effect than the same coding sequence in the male. For example, imprinting plays a role in disorders such as Prader-Willi syndrome (gained through the father) and Angelman syndrome (gained through the mother), but there is increasing evidence that subtle differences in the expression of imprinted genes can contribute to more common disorders like obesity, diabetes, psychiatric illness, and cancer (Peters, J. 2014; also see Carey 2013). Once again, not all epigenetic modifications are beneficial, especially when they are initiated (removed) by outside environmental stressors.

RNA and Junk Genes

Not all DNA/RNA codes for protein; only around 2 percent does. That which does not is often classified as "junk genes" (ncRNA—non-coding RNA) or obsolete genes with no purpose. This has recently been challenged. Prokaryotes (bacteria and archaea) likewise can have lots of ncRNA. However, it appears these so-called Junk genes accumulate with eukaryote complexity. Prokaryotes have less non-coding DNA because they need small "portable" genomes which largely prevent or select against non-coding DNA. Multicellular organisms (eukaryotes) need a far higher level of complexity and subtlety in gene function and this explains, in part, why eukaryotes have much more "junk" DNA than prokaryotes. Some junk genes, however, may be strategies for adaptation; no longer useful, but perhaps a survival strategy marked for the future. These non-coding sequences are like blank checks or phenotypes yet to develop. Recently discovered uses of these ncRNA include regulatory ncRNAs, that adjust bacterial physiology in response to environmental cues, and housekeeping ncRNAs necessary for normal functioning of the cell (see Giles et al. 2016: 117-148; Mills et al. 2016: 183-208).

Noncoding RNAs can base-pair with mRNAs and change their translation efficiency and/or their stability, or they can bind to proteins and modulate their activity; ncRNAs are key regulators. They also interact with the surrounding genomic environment and increase the ability of organisms to evolve by serving as locations for genetic recombination and by providing new and important signals for regulating gene expression. The term "junk DNA" deterred mainstream researchers from studying noncoding genetic material for many years, thinking it was a waste of research time and dollars. This shows us how powerful specific symbolic references like "junk" and "random," delivered by those seemingly in authority, can solidify models, and deter us from others—see Kuhn (1967). RNA comes in many forms and, until recently, again, most were considered junk (Marshall 2017: 272-278), left over from random mutations of the past. RNA comes in many forms—mRNA, rRNA, tRNA, siRNA, micro(mi)RNA, piRNA, ncRNA, etc., and some forms can have direct or indirect effects on coding sequence expression. For example, mRNA, produced in the nucleus, is transported through nuclear pores to the cytoplasm, and functions to bring information from DNA in the nucleus to ribosomes in cytoplasm to direct which polypeptides are assembled. Located in the ribosomes, rRNA is not involved in coding but directs protein synthesis. Transfer or tRNA is located in the cytoplasm, and during polypeptide synthesis, tRNA molecules position each amino acid at the correct place on the polypeptide chain, then tRNA attaches itself to one end of the chain, and transports the amino acids to the cytoplasm. Micro(mi)RNA functions to remove RNA transcripts, as often there are too many of them. The result is less mRNA available to serve as a template for protein. This miRNA-based form of gene regulation, known as RNA interference, functions to fine-tune the amount of protein made for a particular gene. Small Interfering RNA (siRNA) inactivate mRNA after transcription. Result—no translation. neRNA, a non-coding RNA, doesn't code for protein, and, in a similar fashion as piRNAs, is linked to gene silencing of transposons in germ cells, particularly those involved in sperm formation, and it was recently identified in the brain. Below is a list (not all-inclusive) of some classes of RNA involved in epigenetic gene expression:

ncRNA	Noncoding	macroRNA	Macro
rRNA	Ribosomal	sRNA	Small
tRNA	Transfer	snRNA	Small Nuclear
circRNA	Circular	snoRNA	Small nucleolar
eRNA	Enhancer	siRNA	Small interfering
ceRNA	Competing endogen	iRNA	Interfering
lncRNA	Long noncoding	miRNA	Micro
NATs	Natural antisense		
as-lincRNA	Antisense long noncoding		

List of various types of RNA – not all inclusive.

There is also what we call DNA Transposons (discovered by Barbara McClintock, 1950) and Retrotransposons (Transposable Elements), which are segments of DNA that can move around to different positions in the genome of a single cell. In the process, they may create or reverse alterations to the coding sequence. These can accommodate many methyl-groups. Retrotransposons may have originated from retroviruses—I'll have more to say about this in Chapter 4, Symbiosis. Are Transposons caused by random mutations, environmental factors outside the control of the cell, or are alteration caused by the cell, the executor as part of the cell's survival potential? It is important to note that most epigenetic attachments are removed during meiosis, but not all, and those not erased can be passed to the next generation. Epigenetic attachments most likely maintained are those added or removed during gestation because of various stress factors, i.e., nutritional, toxins, emotional stress of mother, and so on, as evidenced initially by the Dutch Hunger Study, discussed below (also see Mill et al. 2016 regrading epigenetic inheritance).

In 2000, the medical community was confident that they could locate "disease genes" and, by altering the genes, develop cures. We have a long way to go as we need to identify enhancers, silencers, etc., that contribute to disease expression. Further, many diseases thought to be mutations, for example, Alzheimer's, color-blindness, Otosclerosis, Crohn's disease, many cancers, type 2 diabetes, lupus, obesity, and many, many others, are tied to epigenetic issues. The addition and removal of methyl units has numerous causes including chemicals in foods, heat stress, cold stress, altitude, social stress, malnutrition, cellular toxicity, viruses, prions, and so on. Here is a list of other epigenetic-influenced diseases, syndromes, and cancers – not all inclusive (see Blakey and Litt 2016; Huang et al. 2016).

Developmental with craniofacial features

Silver-Russell Syndrome (SRS)
Wolf-Hirschhorn Syndrome (WHS)
Developmental/progressive Duchenne

Muscular Dystrophy (DMD)
Hutchinson-Gilford Progeria (HGPS)
Werner Syndrome (WRN)

Complex metabolic/organ disorders/diseases

Chronic kidney disease (CKD)
Cardiovascular Disease (CVD)
Metabolic syndrome (MetS)
Complex neurodegenerative disease Alzheimer's Disease (AD)
Creutzfeldt-Jakob Disease (CJD)
Neural Tube Defects (NTDs)

Developmental with tumor formation

Beckwith-Wiedemann Syndrome (BWS)
Cancers Breast cancer 1, early onset (BRCA1) Breast cancer 2, early onset (BRCA2)
Clear Cell Carcinoma (CCC)
Squamous Cell Carcinoma (SCC)

Disorders/diseases and cancers of the blood

Hereditary Persistence of Fetal Hb (HPFH)
Myelodysplasia Syndrome (MDS)
B-cell Lymphoma (BCL)
Diffuse Large B-cell Lymphoma (DLBCL)
Myeloproliferative Neoplasis (MPN)
Acute Myeloid Leukemia (AML)
Chronic Lymphocytic Leukemia (CLL)
Chronic Myeloid Monocytic Leukemia (CMML)
Promyelocytic Leukemia (PML)
Stem Cell Leukemia (SCL)

I mentioned earlier that the addition or subtraction of methyl units from DNA or acetyl units from histones, and so on, was outside the realm of random mutations, so let me clarify this. Random mutations to cytosine nucleotides remove any attached methyl units. DNA polymerase corrects most of these types of mutations. Not corrected is the reattachment of the methyl unit(s). All the above serious medical conditions are proof positive that random mutations are destructive because they remove necessary information from the system.

Methylation reactions are carried out in our cells by one of three enzymes, DNMT1, DNMT3A, or DNMT3B. DNMT equals DNA-methyltransferase. It is unlikely that random mutations would cause DNA-methyltransferase to replace a methyl unit; most mutations remove, not replace information.

Epigenetics and Stress

Stress generated in an organism during an extinction event, famine, heat stress, cold stress, predator stress, etc., is a trigger for methylation (or acetylation of histones, etc.) accompanied by behavior change, which is then acted on by "natural selection" (my definition of natural selection differs from the Neo-Darwinists and includes endocellular selection prompted by the cell). This is one reason intermediate forms are uncommon—there are few of them; natural selection forces survivors to adapt quickly through endocellular selection—by making choices at the cellular level, which require thinking and "consciousness" or awareness of stressor.

Stress reactions are key to understanding many epigenetic coding sequence modifications. How chemicals are added or subtracted during stress is complex and is under intense investigation. A key, at least for the human animal, is the glucocorticoid receptor sites (Mineralocorticoid receptors–MR) that bind both mineralocorticoids and glucocorticoids and are found in both sodium or Na(+) transporting epithelia (e.g. kidney, colon) and nonepithelial tissues (e.g. heart, brain). The glucocorticoid receptor (GR, GCR, or NR3C1) is the receptor to which cortisol and other glucocorticoids bind. Here is the point: we find the glucocorticoid receptor in almost every cell in the body; it regulates coding sequences governing physiological development, metabolism, and immune response. Because the receptor gene is expressed in several forms, it has many or pleiotropic effects in different parts of the body. When glucocorticoids bind to GR, its primary mechanism of action is the regulation of gene transcription (see Rhen and Cidlowski 2005, and Lu et al. 2006). In humans, the glucocorticoid receptor protein is encoded by NR3C1 coding sequence, on chromosome number 5 or 5q31. Because it regulates coding sequences governing physiological development, metabolism, and immune response, stress reactions and epigenetic additions and subtraction play an important part in signaling when and what alterations need to occur to deal with the prevailing stress or stressors before and after parturition. Steroids floating around in the bloodstream can be very dangerous, especially in *maintaining* a stress reaction, and the more glucocorticoid receptor sites you possess the quicker you can remove these chemicals and reduce stress. As one example, the more glucocorticoid receptor sites, the less prevalent autoimmune diseases.

Epigenetics and Individual and Social Stress

Well known to psychologists, parents who are "nurturing" have children who likewise become nurturing parents. The same applies, apparently, to most mammals. Rat pups groomed and licked frequently by mother rat usually grow up to be calmer and less anxious than do pups licked/groomed less often. Those ignored are more agitated and rarely make good parents. This also applies to humans and cats. There are two general types of cats connected to our extensive feral cat community. Those born outside and only after a month or so are captured, neutered/spayed and then released, and those pregnant females who are trapped and allowed to give birth indoors. We (unfortunately) get attached to the newborn from day one. They always receive lots of attention—cleaning eyes, grooming, veterinary visits, and so on. I will make a generalization—they are always calmer and aren't as skittish as those who didn't receive this amount of attention. This is a message to anyone with or planning to have children. Nurtured children who possess emotional and behavioral responsibility don't go around shooting or hurting others. It would be nice if politicians and school administrators would catch on to this and concentrate on the cause of these socially disruptive behaviors rather than the means to inflict misery on others.

Nutritional Stress

At the end of WWII, Dutch railroad workers went on strike to stop the transport of German troupes to help advancing allied forces. The allied push failed, and the Nazis blocked all food and medicine from entering the country, and so punished the Dutch. After the war researchers conducted a study involving women pregnant at the time and effects on their offspring. Women starved at the end of their pregnancy (last trimester) gave birth to children of smaller birth weight; when these children grew up, their children also had low birth weights. On the other hand, women starved only at the beginning of their pregnancies (first trimester) gave birth to children normal in terms of birth weight, but tended to obesity or being overweight as if overcompensating. They likewise passed this tendency toward food overindulgence on to their children.

Men who experienced famine in childhood were less likely to have grandsons with heart disease or diabetes than those well fed! Grandsons who had lots of food available when they hit puberty often died around 6 years earlier (often of diabetes) than the grandsons of those exposed to famine at the onset of puberty. Starvation seems to trigger the production of "starvation-responsive small RNAs" that are passed down to future generations—the physiology adjusts for famine for the next generation (Roseboom et al. (2001; also see Heijmans et al. 2008). In short, cellular physiology is "thinking ahead." Here we see an example where, while in the womb, the fetus is taking charge of its own development and ability to survive after parturition. These alterations show a preparation for the future by modifying the expression of certain genes involving caloric needs. This cannot be explained as a random mutation without purpose, and it's also of interest that the Neo-Darwinists pay little attention to this research (see Li, et al. 2011; Martincorena et al. 2012; Perkins et al. 2009). Knowing what we now know about fetal involvement in their own development, making choices and adding and removing chemicals from DNA and histones, pretty much blurs our concept of when life begins. Does it simply begin at birth or perhaps life begins at conception? If one believes life begins at birth, then abortions do not take a life. In short, abortions aren't murder, premeditated in most cases. If you believe life begins at conception, as I do, then abortion is murder. Birth control and abortion were promoted in the United States by Margaret Higgins Sanger (1879-1966), a believer in eugenics. What most people don't know is that Sanger was a racist and thought the "inferior" African Americans were having too many babies. There are two problems I have with abortion: 1) the definition of life, or when life begins; and 2) the practice of emotional and personal responsibility. Use of contraceptives to prevent unwanted pregnancy is responsible behavior; birth control via abortion is not. Obviously to conceive it takes two (even with virgin birth), and it is not just the responsibility of women to prevent conception. Sanger, for the most part, left men out of the equation. And who do we call when we want a definition of life? A medical doctor, precisely the individual who performs the abortions; this would surely be a conflict of interest in other economic arenas. Again, when politics enters the debate the motives in favor or against abortion go beyond the rights of women; women, like men, have the right to be responsible, and if you can't be responsible, stay away from the opposite sex.

Longevity and Stress

Telomeres, the distal ends of chromosomes, play a critical role in the control of aging. Telomeres are like the plastic ends of shoelaces. The more a cell divides, the smaller the telomeres become—except in the germ cells that give rise to egg and sperm. Eventually the telomeres get so small they stop dividing and they may activate self-destruct mechanisms. There are coding sequences in the telomeres, but they comprise hundreds of repeats of the sequence TTAGGG. There are no CpG motifs at the telomeres so there can be no DNA methylation. The type of epigenetic modifications normally found at the telomeres and sub-telomeric regions are highly repressive. The epigenetic modification attracts proteins that coat the ends of the chromosomes, help them stay tightly coiled up, and dense and inaccessible. All telomeres have the same DNA sequence—this can be a problem as identical sequences in a nucleus find and bind to one another. By coating the telomeres with repressive modifications, the ends of the chromosomes are densely packed and less chance that different chromosomes will join up inappropriately. Trying to slow down aging, however, presents a problem—without understanding all the factors that go into aging and shrinkage of the telomeres we could speed up aging or create cancer. A recent, and rather shocking study has shown that telomeres can shorten much more rapidly in children exposed to stressful situations (see Mitchell, et al. 2014).

PTSD

Post Trauma Stress Disorder (PTSD) is a condition whereby a person exposed to a traumatic event(s) (war, rape, etc.) continues to experience distressed even when he/she is no longer in danger. Research shows that incidences of stress can lead to transpositionalled changes in the genome. Increased transposon activity was detected in the brain region called the amygdala, which plays a crucial role in emotional responses and decision-making (Riley, et al. 2013). Glucocorticoid receptor sites are at play with PTSD, for apparently traumatic events shut down receptor sites and glucocorticoids continue to circulate; if they are not removed, stress reactions are maintained. The body's purpose here would be to maintain an alert reaction in, for example, a war situation. Maintaining the stress reaction after the danger has passed can be mentally and physically debilitating, and it is unlikely that counseling or "talk" therapy is very useful in these situations. In my private practice dealing with PTSD I found talking and analyzing to be ineffective but found hypnosis useful, especially regressing the patient to a time period before the trauma and then bringing those coding sequences forward in time. As it is the mind that is shutting down the receptor sites perhaps the mind can replace them.

Aging and Cell Death (Apoptosis)

Aging and cell death are natural and occur through a genetically controlled cell suicide process, the purpose of which is to eliminate obsolete or damaged cells in multicellular organisms. Why do we die? What is trying to survive are the coding sequences formally known as genes, and an animal's phenotype is a test of those instructions/codes to determine what works and what does not. There would be absolutely no purpose for testing a phenotype if the cell already knew what worked in a particular environment. So, in a sense, there is direction, there is a "plan" or set of rules, but I do not confine these rules to a narrow slice of the universe, i.e., life forms on planet Earth. Probably life, and the order/rules necessary for that realm, was encoded in the initial nanoseconds of the formation of the universe—or before; certainly, the rules for the formation of hydrogen gas, stars, planets, and so on, were. Anyone who thinks the universe came into existence in some random, haphazard manner misses the complexity and the digital manner in which the inorganic and organic worlds come into existence. This is a step-by-step process, and many of these steps can apparently come together all at once under the right circumstances.

Apoptosis (cell death) is accomplished through dormant enzymes called *caspases*. Coding sequences controlling apoptosis are called reaper, *grim*, *sickle*, and *hid* (scientists have a sense of humor too). Stressors (social stress, malnutrition, toxins, etc.) can enhance aging and cell death and suppress apoptotic genes during tumor (cancer) development. If you think about this for a moment, you might suspect that the cell understands the rules of efficiency and when cells are damaged or suffering from efficiency issues, the cell disposes of them. This means "thinking ahead." In a dualistic world (at least) there is a plus and a minus in terms of entropy: there is life, or the accumulation of energy, and death or loss of energy. Systems wear out, like your car, your air conditioning, and your personal vibrator, and they need replacement. One would expect that different plants and animals have different life spans in most cases related to energy use.

HOX Genes and Developmental Gene Regulatory Networks (dGRNs)

HOX genes (homeotic or "homeodomain" genes) are ancient and necessary for building/signaling the body plan of animals. They are very conservative for a simple reason—they work in terms of the animal's overall physical development and ability to obtain food and engage sex. To move about an animal needs to have limbs, internal organs, etc., in the correct position, in matched opposition or they don't work; random mutations can disrupt *HOX* gene functions. Different body plans require different cells, different proteins, and different support networks. Body plans are modular and require many alterations in coding. Random mutations are just that, random—random mutations would never, statistically, lead to the modular patterns required for each new body plan (McGinnis and Kurziora 1994: 58-66).

Fine-Tuning a System

Humans are bipeds, and upright walking, with a locking out knee joint, is an example of a fine-tuned system. There are many components to upright walking, which not only include a locking out knee joint but a reconfiguration of the spine, pelvis, ankle, and foot. Several of our ancient ancestors suggest there was a gradual fine-tuning of bones, ligaments, and muscles once they realized full commitment to bipedalism. I will have more to say about bipedalism in Chapter 7. For now, however, it should be obvious that random mutations will not selectively fine-tune a system because random mutations are not selective, and to fine-tune a system any changes to a coding sequence have to be *selective*, and *purposeful*. As an analogy, once you invent a functioning gas-powered engine, you can tweak it—you can add weight or reduce weight, increase or decrease the size of the cylinders, cylinders can be in a row or side by side, and so on. We can see these as incremental changes occurring within an already functioning interrelated system, in this case, brought about by the human engineer. The same applies to cells; a certain amount of fine-tuning can occur, changes prompted by the cell/organism, as long as they follow rules. Overtime fine-tuning can lead to new species or possibly extinction, if too specialized. An example of the latter would be the fine-tuning of a digestive system that prefers only one type of food, leaves perhaps from a special tree or bush.

A major problem at this point is establishing patterns in these epigenetic processes, and the frequency of epigenetic information being passed from one generation to the next, escaping the resetting during meiosis that is thought to occur. Epigenetic patterns leading to new species is unlikely to be random.

Epigenetics and Speciation

As Noble (2017: 123), discussing Darwin's finches states:

> In this connection it is interesting to note that the genetic and epigenetic analysis has now been done on the Galapagos finches. The results show that at least as many epigenetic *as genetic changes underlie the differences between the various species and that the number of epigenetic changes correlates rather better with evolutionary distance between the species than do the genetic changes* (Skinner et al., 2014). At the least this result puts the standard explanation for speciation in doubt in this iconic example. From the experimental information alone it would be impossible to say whether epigenetic changes led the speciation with subsequent assimilation into the genetic material or vice versa. Both would be possible. Even more likely, the two naturally go together. (emphasis mine)

One problem involving speciation is the addition of methyl units, etc., to the cytosine bases on the DNA strand and lysine tails of histone would, theoretically, have to occur on the sex chromosomes if an alteration is passed on to the next generation. As any organism is a complex network of interacting cells, we should not exclude communication with the germ cells. I don't think we know enough about cellular signaling systems to say for sure if this is the case. Zenil, et al. (2017:273) state:

> *Important sources of information are epigenetic phenomena, an additional layer of complexity reversing the traditional molecular biological dogma that describes how information is transferred from the genome all the way to the upper levels. Epigenetics shows that information can flow bottom-down from all upper layers to the lower layer (genome). This information coming back to the cells alters how genes are finally expressed even if the cell's genome (the DNA code) remains exactly the same. Epigenetics is the process in which external information is introduced to the various cell layers that can disrupt the function of it and lead to changes, including diseases.* (emphasis mine)

Epigenetic changes and species isolation/separation for long periods of time can lead to new species, keeping in mind that other alterations combining with epigenetic additions and subtractions would likewise result in new species. Complimentary types of alterations could include tandem repeats, deletions, insertions, duplications, translocations, inversions, conversions, and mitochondrial alterations. An interesting comment about duplication and speciation is as follows: duplication comprises a piece of DNA that is abnormally copied one or more times. This type of alteration may modify the function of the resulting protein. A sequence connected to the cortical part of brain in the genome of humans is duplicated *212* times as compared to *37* times in chimps (Mitchell and Silver 2018)— although these repeats might have come from decisions made by the cell, duplication seems to be connected to viral control (see Chapter 4, Symbiosis). What is the probability of random mutations affecting the same coding sequences(s) causing 212 repeats? This is an incomprehensibly large number, way beyond any reasonable implication these repeats were caused randomly. Keep also in mind that the cell, the executor, is not a passive recipient of what nature provides but an active participant in its own survival. Even Dawkins (2006: 29) agrees with this:

> The evolutionary importance of the fact that genes control embryonic development is this: it means that genes are at least partly responsible for their own survival in the future, because their survival depends on the efficiency of the bodies in which they live and which they helped to build.

The last part, "responsible for their own survival in the future," serves to undercut his position that coding sequences are passive recipients of random events, along with life without meaning or purpose. But Dawkins then takes it all away (2006: 68):

> The genes to control the behavior of their survival machines, not directly with their fingers on puppet strings, but indirectly like the computer programmer. All they can do is set it up beforehand; then the survival machine is on its own, and the genes can only sit passively inside. Why are they so passive? Why don't they grab the reins and take charge from moment to moment? The answer is that they cannot because of time-lag problems.

At issue is understanding what is making the survival decisions? It isn't the environment—the environment is only providing the stressors, the information. Is Dawkins suggesting pre-programming, and if so, where did this pre-programming come from? What Dawkins didn't anticipate is the dynamics of the executor, the cell, and its active involvement in its own survival through "thought" and decision making.

Speciation evolving out of epigenetic—especially cell-based decision making and choices, is best illustrated through splitting of groups, for example our hominin ancestors, during the food quest. As populations increase, the environment, in most cases for hunters and gatherers, cannot provide sufficient calories. This creates stress and forces groups to split and move away from one another, to seek greener pastures. We are dealing with small groups that separated in time and space and this can lead to inbreeding and the accentuation of negative coding sequences, thus undercutting the health of the group by corrupting or removing information from the system that could be useful in the future. Sanford (2014, and Rupe and Sanford 2017) and others suggest that such inbreeding was likely one reason many clads went extinct. Modern humans (after c. 300 KYA) would have experienced the same issues of inbreeding. However, new information (see Chapter 7) suggests modern human mutation rates have slowed, at least since our assumed split with the supposed common ancestor to chimps and ourselves. The other side of this involves the reuniting of the groups thousands of years later; reuniting produces hybrids often more fit than the parents. Another issue is, how many of the mutations, even ones considered neutral, are directed/initiated by the cell as a response to environmental stressors and not random at all? Let's face it, the cell is smarter/more knowledgeable than modern humans will ever be about its past and future. We have to accept the fact that thinking and decision making is not the sole provenience of *Homo sapiens*, just as we slowly accepted the fact that planet Earth is not the center of the universe. Back to epigenetics.

Groups split, and as time goes by, there are cell directed epigenetic additions and subtractions, there are spelling mistakes, there are the Neo-Darwinian mutations, plus the other alterations (insertions, duplications, etc.) mentioned above in response to environmental stressors. This pushes the genotypes and their phenotype expressions further and further away from the survivors of the original group. It takes approximately 2 million years of separation to produce a "new" species incapable of interbreeding and producing offspring with the original group members. Keep in mind that evolution is always *away* from some specific form, *not toward* any specific form, and this non-specificity (non-predetermination) is a survival strategy, a continual process for stemming entropy. That's what life is, at the minimum—a negative entropy strategy to, perhaps, afford enough time for intelligent beings to figure all this out. The question remains: how many of these epigenetic alterations are cell directed as opposed to random mutations? Combine this with tandem repeats, deletions, insertions, duplications, translocations, inversions, conversions, and mitochondrial alterations, the process of speciation could proceed rapidly, certainly in less than 2 million years; thought and decision making might help to explain the Cambrian explosion and the lack of intermediate forms.

Another possibility is through genotypic and phenotypic coding altered or enhanced by symbiosis and hybridization leading to rapid speciation (see Chapters 4 and 5; also see Bermudes and Beck 1991). In other word, there may have been a great deal of informational exchange between life forms in the early Cambrian, leading to new species and phyla and, because this occurred rapidly, it only appears that all the new phyla showed up at once.

Thinking/Decision Making at the Cellular and Quantum Levels

Escherichia coli (*E. coli*–a prokaryote) is a gut bacterium with many strains, some of which are deadly. However, *E. coli* can do a neat "trick"—it can "flick" a switch and go from consuming glucose to digesting lactose (lac operon). This is like a "frame shift," and brings us to decision making at the cellular level and more probably "thinking" that occurs at the quantum level in reaction to environmental stressors. Thinking and decision making are not the sole provenience of humans. Decision making was possibly imbedded in the origins of the universe, just as the ability for cells to combine leading to self-replicating life forms. This thinking and decision making may be part of the puzzle relating to all the new phyla, without fossil intermediaries, as mentioned above, connected to the Cambrian explosion of c. 550 MYA. Perhaps we just haven't found the intermediates as yet, but they should be visible in the strata that precedes the Cambrian—no such luck. What we see is rapid, substantial evolutionary change with shifts into new adaptive zones—again, RAPIDLY. Once a threshold is reached in acquisition of a new adaptation, strong directional selection, mediated by the cell, shapes features (epigenetics) into new forms. The term I use for this *morphological momentum*.

Diauxic Shift in Yeast (Eukaryotes)

I similar situation occurs in yeast. Yeast cells metabolize glucose primarily by glycolysis, producing ethanol and carbon dioxide. Those who brew their own wine or beer are familiar with the process. When glucose is depleted, the cells enter "diauxic shift," or a switch in metabolism from using glucose to aerobic utilization of ethanol (Galdieri et al. 2010). How are yeast cells able to do this without "thinking" or planning? How did this option occur? It could not possibly have come about by random mutations, as "random" is not specific enough to bring about such a process. Simply put, yeast cells are hungry, and when the food supply runs out, they go dormant. The ability to shift from glucose to ethanol would likewise enhance survival, and, in my opinion, yeast gained this ability during the dormant phase. Both alterations (*E. coli* and yeast) occur through the addition or subtraction of chemicals from DNA and histones.

Metamorphosis and the Salamander (Axolotls)

Shubin (2020: 41-45) presents an interesting case of epigenetics (although he doesn't use this term) for the metamorphosis of a Mexican salamander, Axolotl. This creature can exist in two environments, one wet and one dry. "Those reared in wet environments never undergo metamorphosis and grow to look like big aquatic larvae, with a full set of gills, a flipper-like tail, and a wide skull best suited for feeding in water." If the offspring of Axolotls find themselves in a dry environment they lose all these interrelated traits and take on a terrestrial lifestyle. This ability to make such drastic changes to its physiology is not caused by random mutations but, instead, a survival strategy – more complex to be sure, similar to the bacteria and yeast discussed earlier. The changes occurring in Axolotls are alterations in expression of coding sequences rather than altering base pairs or the order of a sequence.

Genetic Dormancy

Humans experience a period of dormancy when we sleep or take cat naps. Dreaming, as we will encounter in Chapter 6, may be an adventure in the wave form, and in the wave form there are many possibilities. We have all heard the statement, "sleep on it," and many times solutions come to us through dreams. After catastrophic events, is it possible that under special conditions coding sequences can go dormant and reconfigure themselves at the quantum level? At the quantum level systems can assemble incrementally because they don't have to function immediately. If this occurs, and this is speculation, this might also explain why there are few intermediate forms. One of the doubts that plagued the early Darwin supporters was how to explain the sudden appearance of all the great animals at the beginning of the Cambrian Era (c. 550–485.5). Except for "sponge-like" remnants and possible tube worms, different forms of trilobites, mollusks, worms, and sponges just show up. Darwin had faith that precedents to the Cambrian explosion would eventually materialize; few have. What would explain this? Some scientists suggest that most of the pre-Cambrian animals were soft bodied and wouldn't preserve well. We've been able to recover fossilized bacteria dating to 3.5 BYA, so I'm not sure the argument is valid. But there is another consideration—what is to follow is speculation. What if DNA and all the support systems can enter a dormant phase? By this I mean, under unusual circumstances (e.g., catastrophic events), where basic life-sustaining structure/systems could go dormant, enter a wave form, and start planning and rearranging for the future. There wouldn't be a problem for irreducible complexity because the DNA and support systems will not manifest in physical form until the crisis is over. What is happening on planet Earth 2.5 BYA until .550 BYA? From approximately 710 MYA until 550 MYA, scientists have detected what we have called "Snowball Earth" (see Lane 2015). There is no quick answer for cause, although the production of a vast amount of oxygen through photosynthesizing organisms (bacteria and eukaryotic forms) has been suggested. The oxygen displaces CO_2, heat more rapidly dissipates, and, combined with radiation cooling, we end up with Snowball Earth. There are other explanations and probably it was a combination of events. The point is this would be a catastrophic event, and cold stress is at the top of the list of factors involved in the alteration of the epigenetic coding sequences and changes in phenotypic features along with genotypic features. Let's take this further. There have been numerous catastrophic events in Earth's history, i.e., the ones mentioned above, Snowball Earth, the Great Extinction of 252 MYA, and the 65 MYA K/T boundary extinction and the end of the dinosaurs. And then there are less global extinctions, mini-ice ages, volcanic eruptions, and so on, that would affect local populations. These events might have triggered genetic dormancy (like sleep, a visit to the wave form), alterations in the coding sequences, and the rise of new species. Keep in mind that there are few (if any, depending on interpretation) intermediate species found before the geologic Eras, Periods, and Epochs, yet afterward there are new animals and plants with vastly different phenotypes which seem too fit, too "coincidentally" selected for these new environments. Let's not forget the ability for cells to think and make decisions does not necessarily suggest they are "controlled" by a super intelligence any more than it necessarily follows that us humans, because we think and plan, are controlled by some supernatural power.

What kind of general phenotypic changes could one imagine especially connected to the Cambrian explosion? Perhaps getting larger because of predatory activities in the late pre-Cambrian time period. Mobility factors, that is, converting rudimentary appendages into fins or limbs, perhaps thin membranous cell walls mineralized to produce exoskeletons, better camouflage, perhaps offensive structures, and so on. By adding and/or removing chemical units from DNA coding sequences and histones (with multi-celled eukaryotes), the organism could accentuate or decrease expression of any system or module. Again, the suggestion is, alterations of life forms, existing in a dormant state, could happen in the wave form with new patterns, perhaps even tested at the quantum level. This is only speculation, but I favor my speculation over that of the Neo-Darwinists'.

Epigenetics and Physical and Mental Health

Putting speculation aside, DNA, including all the support systems, is conservative and its evolutionary development extends back in time perhaps 10 billion years or more. The rules of life are complex, too complex to have evolved through random mutations. Random mutations, that is, changes in the nucleotides/base pairs act as noise in the system and usually lead to zygotes not forming naturally, early death or extreme health issue in life, or decrease in vitality of the organism. On the other side, if a change in the coding sequence is neutral it could result in a preadaptation useful for a future generation. We can see DNA as the "hard drive" with a control panel that accommodates methylation and influences from histones, various enzymes, etc.; these influences are called epigenetic and their design is to keep systems in balance. This servo-system can be knocked out of balance through social stress, poor nutrition, inadequate detox, chemicals in environment (and in foods), viruses, bacteria, etc., and can affect both mental and physical health.

Two major considerations for health are proper nutrition and cellular detox, but there are also genetic predispositions to health and various types of stressors.

Epigenetics, Cellular Toxicity, and Evolution

Phenylthiocarbamide (PTC), also known as phenylthiourea (PTU), is an organosulfur thiourea containing a phenyl ring. It has an unusual property—it either tastes very bitter or is virtually tasteless, depending on the genetic makeup of the taster. The PTC coding sequence (TAS2R38), and the epigenetic variations in the coding sequence, was discovered in 2003. Found in Brussels Sprouts and lesser amounts in cabbage and cauliflower, PTC can be tasted, or not, and those who can taste PTC usually experience it as bitter. This chemical can alter thyroid functioning, and the ability to taste this substance is a warning. This is just one food chemical among many that cause alterations in CpG methylation.

PTC also inhibits the production of melanocytes and is used to grow transparent fish. Melanocytes produce melanin, a pigment found in the skin, eyes, hair, nasal cavity, and inner ear. One has to stop and wonder what environmental and cultural factors led to the development of lighter skin color as human populations moved into more northern climates? Simple random mutations, plant chemicals and methylation, etc.? Perhaps the cells were smart enough to signal changes, or did we pick up the alteration of the melanocytes from the Neanderthals living in the Levant 55 KYA? Were there viruses and bacteria involved? We know that the melanocytes alter when modern humans leave Africa and move into northern and northeastern climates where sun exposure decreases and clothing becomes a necessity. Sunlight is necessary for converting cholesterol, the fat layer under your skin (a peculiar condition for humans but no other primate), into vitamin D3 necessary for the absorption of calcium. (Vitamin D3 serves many functions besides calcium absorption.) When wearing clothes, you cut down on skin exposure and that which is exposed has to do more work—lighter skin allows more production of D3. The specificity here suggests thinking at the cellular level. Without adequate calcium intake, young people end up with rickets; adults get hypocalcemia.

Epigenetics and Cellular Toxicity: More Examples

Children may suffer delayed language skills if their mothers come in contact with common chemicals in many products, including nail polish, hair spray, food packaging, vinyl flooring, etc., while pregnant. The chemical, "phthalate," is found in countless products as plasticizers (make things pliable), and, as solvents, they enable other substances to dissolve. Research reveals the risk for language delay at about age three years was up to 30 percent higher among children whose mothers had higher exposure to two phthalates, dibutyl phthalate-

DBP and butyl benzyl phthalate–BBP. Both chemicals are in older vinyl flooring, cosmetics, and plastic toys. Phthalates are hormonally active and affect the body's hormone system. We know DBP and BBP lower testosterone in the mother during early pregnancy, which helps explain how they affect intellectual development. Scientists have linked phthalates to developmental delays, lower IQ, and under developed male sex organs (see *Newsmax Health* 2018. "Common Chemical Causes Language Delay in Children." November 23).

Epigenetics and Cellular Toxicity—Estrogen in Foods

Many studies show specific hormones in human and animal milk are essential for infant growth and immunity. Increasing evidence indicates environmentally (and socially) introduced hormones in dairy products, skin products, birth control pills, etc., impact human health including the role of some estrogens and insulin-like growth factor-1 in initiation of breast enlargement, breast, prostate, and endometrial tumors, and alterations in sexual orientation (see Hines 2011). These health issues are not a "mutation" of nucleotides or base pairs, but the increase or decrease in the expression of coding sequences through the addition or subtraction of chemicals from DNA and histones because of cellular misdirection caused by estrogen or chemicals that imitate estrogen.

Pharmaceutical/prescription drugs can also cause misdirection and lead to severe injuries and death; these are termed *iatrogenic*, which means caused by medical examination and/or treatment. My advice to anyone who regularly sees Western medical doctors, if there is no clear diagnosis, don't accept treatment, and always, always seek a second or third opinion; misdiagnosis is very high.

Epigenetics and Its Impact

The dogma presented by the Neo-Darwinists, i.e., "evolution can only occur with the alteration/mutation of base pairs or their sequence," is a false assumption stated and restated as "fact," without scientific validation, for many decades. This opinion, "validated" by authority rather than scientific data and fact, has held up our understanding of life's processes. Genetics is much more complication and not as straightforward as the Neo-Darwinists would have it. Regarding health, malnutrition and cellular toxicity are the major causes of disease, not germs and bad genes. Malnutrition and cellular toxicity come from life-styles that promote fast foods and lots of sugar, fat, and salt, and from a medical system that pays little attention to nutrition. We know more about the nutritional needs of a chicken than we do of a human being! Nutrition not only involves the foods we eat but when and how we eat. We have all heard of reflux. The medical community, along with the pharmaceutical industry, have a cure—drugs that stop the production on HCl (hydrochloric acid) and other digestive enzymes; if you can't digest your food, there is no gastric distress. Now you can eat all you want, when you want. What a relief, right? Indigestion is a symptom, an "inner voice" telling you something is wrong. And because your stomach hurts and not your foot, it's probably because of something you placed in your stomach; it isn't because you stubbed your toe. Indigestion-designed drugs do not allow the proper digestion of food and gradually organs suffer; your liver is working overtime, not only detoxifying the garbage food you are eating but the dangerous pharmaceutical drugs you are taking. There is a simple cure for reflux; always wait three hours after having a meal before you lie down and go to sleep. Why? When you swallow food, it passes into the esophagus moving first past the upper esophageal sphincter (valve) along the esophagus and then it passes through the lower esophageal sphincter before emptying into the stomach. When you go to sleep on a full stomach, great pressure in placed on the lower esophageal sphincter, and over time this weakens the valve allowing partially digested food and HCl to enter the esophagus and erode the tissue. Also, don't eat gut buster meals at dinner time (or any time) as it takes longer for food to digest; small meals are more easily digested.

Further, there are only two types of food—dense animal protein (eggs, beef, pork, chicken, fish, etc.), and everything else. Never mix dense animal protein with anything else because you cannot properly digest the protein. For the reasons why, consult a clinical chemistry book. The medical community, in concert with the USDA, instruct people to eat the way we *culturally* eat, not in a way that matches your digestive system.

Hunter/gatherers do not have the luxury of a smorgasbord; they eat whatever they find along the way, and no disrespect to the USDA, finding all the food groups in one place at the same time would have seemed miraculous. Our guts/digestive systems evolve/change just like everything else, and I suspect that 20 million years of fruits, berries, tubers, and veggies would have a profound effect on how and what we should consume today. The medical community will have to pay more attention to clinical chemistry and nutrition. However, nutrition can't be boiled down to "one size fits all."

Psychiatry/Psychology

Extreme stress/trauma (both physical and emotional), malnutrition, cellular toxicity, and lifestyle can alter coding sequence expression leading to emotional/behavioral issues. Many of the "talk" therapies are ineffective in resolving these issues because of the actual chemical changes in the body. Talk therapy, regardless of the interpretative model you subscribe to, is usually designed around the mistaken assumption that by talking the patient will talk out his or her problem, new solutions will emerge, and the issue will go away. The therapist will offer interpretations along the way, but this is not scientific, where there is a clear cause and effect. In fact, if allowed to talk, one inadvertently anchors the story further and further into the psyche as truth. The patient is having trouble with the narrative, the story, and the best way out is to change the story, the symbols of distress. You can't change history, but you can change the way the individual interprets his or her history. A patient's beliefs about history are never "true" in the sense of being accurate, and the further away the individual gets from a specific historical event or trauma, the more distorted the narrative becomes. Why? The mind (and obviously the brain) is attempting to reduce stress by removing or adding information; the therapist's job is to help build a story the patient can live with. Drugging people in most cases is not the answer either, although, in the future, chemistry might be useful in those cases where the addition or subtraction of methyl units, etc., would offer an actual cure rather than suppression of symptoms.

Social Anthropology/Sociology

The nature vs nurture argument is settled; both are involved in the enculturation process. The propensity toward violence, for example, is mirrored in models presented by one's cultures, but there are likewise genetic issues involved in terms of predisposition to violence (Garcia-Arocena 2015). Social factors including stress, substance abuse, diet, sleep quality, and social relationships which we face throughout life, all contribute to shaping our brain functions by the addition to, or subtraction of chemicals from DNA and histones. Changes in the expression of specific coding sequences in the brain, such as MAOA, DAT1 and DRS2, can affect neurotransmitter levels, which impact intelligence, mood, and memory. The MAOA coding sequence (type A monoamine oxidase), on the X chromosome, is also known as the warrior gene, since abnormal versions of the gene often result in aggressive behaviors. MAOA's function affects the following systems: the dopamine system is connected to mood, motivation and reward, arousal, and other behaviors; the serotonin system, which is involved in impulse control, affects regulation, sleep, and appetite; and the epinephrine/norepinephrine system, which facilitates fight-or-flight reactions and autonomic nervous system activity. Ethyl alcohol (ethanol), the drug in beer, wine, vodka, and other alcoholic beverages, is a monoamine oxidase inhibitor and accentuates the MAOA coding sequences, linking alcohol use/abuse to, not only violence, but the removal of all social rules.

At one time social scientists believed humans are born with a "blank slate," often incorrectly attributed to John Locke (1632-1704), also considered the "father of Liberalism." Some Neo-Darwinists and sociologists still subscribe to this belief. We now know that humans have many predispositions, for example, leadership, deception, athletic abilities, art forms, aggression, and so on. Predispositions to behavior involve identifiable coding sequences and, just like everything else coded within the genetic material, are subject to change from group to group due to long periods of separation from an original group. Thus, not only are their individual predispositions, this should likewise manifest itself in groups. This is controversial as we would like to think all humans are on the same wavelength regarding predispositions, but this is unlikely to be the case (see Wade 2014; Murray 2020). Sociologists and social anthropologists who are attempting to understand human/cultural behavior have to take genomics into consideration. Social systems emerge from our biology.

Unexpected Critique of Epigenetics

Leisola and Witt (2018: 207-211) consider epigenetics (along with the theories of self-organization, Evo-Devo or evolutionary development biology, natural genetic engineering discussed in later chapters), as hollow fixes to the Neo-Darwinian position. They state:

> These are some of the more prominent patch theories put forward in response to experimental data but also in the hope of rescuing modern evolutionary theory. There are many others including hybrids... As naturalists, the researchers pushing these hypotheses do not doubt evolution as such. For many of them that is not allowed. They only doubt the neo-Darwinian mechanism. But all of the proposed patches come with one or more fatal limitations—and in every case the limitations can be boiled down to an inability to generate novel and advantageous biological form and function.

Leisola and Witt seem to suggest that epigenetics, hybridization, and so on are useless research avenues and should be abandoned because the data, at the moment, does not come to the Intelligent

Design position. This is little different from the Neo-Darwinists pushing ID aside. The definition of mutation by both the ID and Neo-Darwin groups is much too narrow. Evolution refers to change—slow or fast, and not that it is necessary for one species to "change" into another. Symbiosis is a good example of fruitful research where "novel" information is created. With symbiosis you can arrive at a new species by creating a new separate phylum or family, rather than climbing some imagined tree. This may not be new information, in the sense of never having existed on planet Earth, but when combined with other information leads to essentially a new species or adaptive potential. There is recent research showing that a virus invaded (interacted with) *HOX* coding sequences of early mammals leading to the development of the placenta and initiating the Eutheria or Placentalia line of mammals (see Chapter 4 and 5). Recall the Neo-Darwinian proposal of random mutations accidentally creating species; the research cited by Leisola and Witt serves to disprove the Neo-Darwinian model.

Lateral Gene Transfer

Lateral gene transfer is considered by some (see Ward 2018: 101-104; also see Pal et al. 2005) to be an epigenetic phenomenon, and, in terms of "adding" or "subtracting" chemicals from the genetic code, new information is added through the transfer of these sequences from one organism to another (also see Chapter 4). Ward comments (2018: 101):

> In lateral gene transfer (LHT), large portions of the DNA of one archaea or bacteria are replaced through addition from a second microbe, and one of the exciting discoveries of the past decade is that LGT is not just restricted to the Prokaryotes. Even humans have been evolutionarily changed by the sudden addition of new genes, injected into us by microbial vectors.

Ward goes on to say (2018: 101-102):

> This is no slow, mutation-by-mutation change. As it does now, it would have taken place probably in a short period of time—

in a few hours, in fact, the LGT process is among the most Lamarckian of evolutionary processes known. Yet, the invaded microbe acquires not only a new trait but whole suites of traits. If discussed in terms of the tree of life, we are not talking about a new twig. LGT sometimes brought about so much novelty that it would have caused the new microbe, because of how new it would have been, to jump to separate families, if not become a new species. The process will also play a large role in the future of evolution on earth, with the vector being humanity as we add new genes to plants and animals that serve as food, or in our attempts to kill weeds or insect pests in crop settings. Again, lateral or horizontal coding sequence transfer occurs when genetic material/information is passed sideways to another, usually unrelated organism, rather than vertically (parent to child). But the extent to which this happens in eukaryotes is difficult to measure because so few eukaryote genomes have been studied as compared to prokaryotes or bacteria.

LGT, make no mistake, is not always positive (see Gillings 2016). The point is that LGT is another avenue for developing new species. Neo-Darwinians see this as another part of their random mutation narrative – this is highly unlikely. Perhaps prokaryotes and eukaryotes seek out these avenues as part of their own survival mechanism; interspecies co-operation or communication is common in the animal world as I will discuss in Chapter 4. One final comment. Gontier (2015: 147-148), referring to antibiotic resistance in bacteria, states:

> The transmission of antibiotic resistant genes via bacterial conjugation shows how rapid evolution through HGT (horizontal gene transfer) can be. In less than 10 years after the first worldwide administrations of antibiotics, Shigella bacteria were able to acquire and spread that resistance. Some of these resistant genes already existed within the bacteria involved, and other resistant genes have evolved since the massive introduction of antibiotics. The current standard paradigm assumes that genetic

mutations are "random," and it considers such random mutations to result from "copying errors." Whether or not the evolution of resistance genes is "random" or "directed" remains a topic of considerable research, *but the rapidness by which transfer occurs across species outnumbers any suspicion of randomness.* (emphasis mine)

The Sugar Code

Until recently the assumption was that DNA, including the coding sequences and chemical additions (epigenetic), are the main information senders and receivers in terms of cell and system management. That is to say, the arrangement of the nucleotides on the coding sequences carries information that prompts cells to function. We often think of sugar or carbohydrates as sources of energy through glycolysis, as part of the phosphate/monosaccharide molecule making up the ladder of the double helix and bond for the nucleotides, or as ribose sugar in ribonucleic acid (RNA) or deoxyribonucleic acid (DNA). Research since the 1990s has revealed that sugar molecules play a much bigger role in cellular metabolism and are significantly more active in the sending and receiving of information than DNA and RNA. A substantial part of the genome is represented by coding sequences producing proteins involved in carbohydrate production, activation, and transport. There are also coding sequences involved in glycan assembly, modification, remodeling, and degradation. Because of the large devotion to carbohydrate production, the importance here seems to surpass that of protein synthesis. Monosaccharides exceed nucleotides or amino acids in coding capacity and, as a result, the glycome is very complex and difficult to study (Laine 1997; Rüdiger and Gabius 2009). Thus, the sugar code refers to researching the coding capacity and the functions monosaccharides play in cellular communication and overall systems functioning. Accordingly, monosaccharides are considered the third alphabet of life, with nucleotides and amino acids being the first and second. Before reviewing the sugar code specifically, I want to share an abstract of a recent article written by Marcello Barbieri (2018:1):

> Various independent discoveries have shown that *many organic codes exist in living systems*, and this implies that they came into being during the history of life and contributed to that history. The genetic code appeared in a population of primitive systems that has been referred to as the common ancestor, and it has been proposed that three distinct signal processing codes gave origin to the three primary kingdoms of Archaea, Bacteria and Eukarya. After the genetic code and the signal processing codes, on the other hand, only the ancestors of the eukaryotes continued to explore the coding space and *gave origin to splicing codes, histone code, tubulin code, compartment codes and many others*. A first theoretical consequence of this historical fact is the idea that the Eukarya became increasingly more complex because they maintained the potential to bring new organic codes into existence. A second theoretical consequence comes from the fact that the evolution of the individual rules of a code can take an extremely long time, *but the origin of a new organic code corresponds to the appearance of a complete set of rules, and from a geological point of view this amounts to a sudden event.* The great discontinuities of the history of life, in other words, can be explained as the result of the appearance of new codes. A third theoretical consequence comes from the fact that the organic codes have been highly conserved in evolution, which shows that *they are the great invariants of life*, the sole entities that have gone intact through billions of years while everything else has changed. This tells us that the organic codes are fundamental components of life and their study—the new research field of Code Biology—is destined to become an increasingly relevant part of the life sciences. (emphasis mine)

What Barbieri is saying is that the coding for life is complicated by the numerous organic codes (DNA, RNA, mono-saccharides, etc.)

necessary for maintenance, housekeeping, and development. What is of equal importance is these codes, and the rules that support them, just show up, without precedents. I have not encountered anything in the Neo-Darwinist's literature that accounts for this, as they are stuck on random nucleotide mutations as an explanation for origins and speciation. In short, it is difficult to discount irreducible complexity, certainly regarding the production of proteins, cells, and so on. We can say the same of monosaccharides, as their communication potential is absolutely necessary for initiating, supporting, and maintaining cell metabolism. And finally, the conservative nature of these codes suggests that they are ancient and have been there from the beginning of life's appearance. Returning to the sugar code, Gabius (2018) comments:

> The cell surface is the platform for presentation of biochemical signals that are required for intercellular communication. Their profile necessarily needs to be responsive to internal and external factors in a highly dynamic manner. The structural features of the signals must meet the criterion of high-density information coding in a minimum of space. Thus, only biomolecules that can generate many different oligomers ('words') from few building blocks ('letters') qualify to meet this challenge. Examining the respective properties of common bio-compounds that form natural oligo- and polymers comparatively, starting with nucleotides and amino acids (the first and second alphabets of life), comes up with sugars as clear frontrunner. The enzymatic machinery for the bio-synthesis of sugar chains can indeed link monosaccharides, the letters of the third alphabet of life, in a manner to reach an unsurpassed number of oligomers (complex carbohydrates or glycans). Fittingly, the resulting glycome of a cell can be likened to a fingerprint. Conjugates of glycans with proteins and sphingolipids (glycoproteins and glycolipids) are ubiquitous in Nature. This implies a broad (patho)physiologic significance.

The alphabet of the sugar code or language is represented by nine right-handed monosaccharides and four left (Rüdiger and Gabius 2009). This is not to be confused with D and L related to amino acids. However, like the D, L convention for amino acids, D and L are used to denote specific shapes formed by these complex compounds. And this is one of the research objectives, that is, to uncover the configurations (glycomic profiles) of these sugars and their capacity to send and receive information. Mammal physiology uses nine different right-handed sugars, along with the 20 left-handed amino acids, amounting to multivalent (many points of attachment) carbohydrate-protein/carbohydrate-lipid associations critical to the development and maintenance of life (Chabre and Roy 2009: 53). The sugar code begins with eukaryote cells and the assumption that eukaryote complexity demands biochemical signals of high coding capacity—monosaccharides are especially equipped for this purpose. Chabre and Roy (2009: 55) comment:

> We now understand multivalent interactions to be a ubiquitous strategy that has evolved in nature for a wide range of functions, that provide numerous and unique benefits that are not achievable with monovalent interactions... Indeed, saccharides are expressed on the majority of mammalian cell surfaces and are bound to proteins ('glycoproteins') and lipids ('glycolipids') that are entangled in the cell membranes and clustered in multi-antennary (antenna or dendrites) configurations.

For those interested in the biochemistry and methods for isolating and identifying the structures of these sugars and their interactions, I suggest the reader consult original sources (see Gabius 2009) and more recent discussions on the internet. Here I want to concentrate on the functions of the sugars on cell and system metabolism.

The Life-Sustaining Activities of Sugar

As mentioned above, besides providing sources of energy through glycolysis, monosaccharides are part of the phosphate/monosaccharide molecule making up the ladder of the double helix and bond for the nucleotides, or as ribose sugar in ribonucleic acid (RNA) or deoxyribonucleic acid (DNA). Beyond these functions, glycoproteins are necessary for the correct folding of proteins into the precise functional shape (Zuber and Roth 2009: 92-96); as "a first line of defense that pathogens have to overcome for cell entry" and to initiate human killer cells (Zuber and Roth 2009: 105-106); sugars are actively involved in decision making for maintaining proper metabolic functioning of the cell (Zuber and Roth 2009:96, 107); "[i]n humans, N-glycosylation is directly involved in the pathogenesis of several diseases such as protein-folding diseases, congenital diseases of glycosylation and cancer" (Zuber and Roth 2009: 109); "O-Man glycans represent a smaller group of glycoprotein modification, and are present in only a few examples limited to brain, neural tissue and skeletal muscle where they play an important role in development" (Patsos and Corfield 2009: 120); "[i]mmune cell expression and response is strongly linked with O-glycosylation in immune system and is apparent at several levels" Patsos and Corfield 2009: 127); cell growth and proliferation, stress (e.g. heat shock), survival, and the cell cycle; disease (e.g., insulin signaling/regulation of glucose toxicity, Alzheimer's disease), protein structure and stability, on and on (Patsos and Corfield 2009: 129-135).

The Sugar Code and Medicine

Many diseases are caused by the disruption of glycosylation, for example, diabetes and Alzheimer's (see Hennet 2009, Chang et al. 2018). This is of interest to the medical community for, in the future, it might be possible to "fix" or repair those disruptions. On the other side, in terms of producing fungicides and antiparasitic compounds (Shams-Eldin, Debierre-Grockiego, and Schwarz 2009), learning how to disrupt the many steps in their biosynthesis would have health benefits as well.

Conclusion

Referenced as the third alphabet of life, with nucleotides and amino acids being the first and second, certainly does not imply sugars are the least important. Eukaryote cells represent a very large jump beyond bacteria and archaea; this Type Two symbiotic relationship (see Chapter 4) between bacteria and archaea produced a cell way more complex than the sum of both, and with that complexity more communication channels are necessary. I have great difficulty believing all this coding, and information sending and receiving could arise out of random mutations, the Neo-Darwinist's magical helpers. Not only is this highly improbable, it is unreasonable. A reasonable possibility is that cells and life forms are more in charge of their change/evolutionary potential than previously thought; epigenetics certainly points this out. The Neo-Darwinists think epigenetics is an insignificant issue in the development of new species. As Dawkins comments (2017: 221-222):

> There is an extension to epigenetics which is more controversial. This is the idea that the pattern of gene use can be passed on to future generations: epigenetic inheritance. We are feted with stories of characteristics being passed on from parent to offspring, in a modern resurgence of the Lamarckian idea of a blacksmith passing on his strong muscles to his children. There seems to be something in the human psyche to which this disastrous concept appeals – disastrous because it would also mean the blacksmith's child inheriting his father's gammy leg, scarred face and political attitudes…

It is obvious from Dawkins' remark that he knows very little about epigenetics as nowhere in modern research has any scientist suggested inheritance of a "scarred face" and "gammy leg." And, yes, the blacksmith will pass on his muscles to his children, and their ability to enlarge, and to what degree through exercise *may* be epigenetic. At this point in research we do not know the extent of communication between the different tissues – including germ cell, to know for sure. Recall the Dutch Hunger Study wherein epigenetic alterations occurring during gestation are passed on to the next generation. Also recall that the cell is the executor, making decisions in response to both external and internal stressors. Nowhere does Dawkins acknowledge this decision-making ability. Dawkins is the master of "Hound Dog Science" (see Chapter 6 for the definition). To discount this possibility is not the way of science. This is a literary technique utilized by Dawkins and others to discredit models contrary to random mutation and natural selection. I'm not sure why he mentions political attitudes for obviously they are passed on through culture. This is sarcasm designed to minimize rather than scientifically confront new ideas; Hound Dog Science is likewise sarcasm – you get what you give. In the next chapter, I consider symbiosis and its place in the evolution of life.

CHAPTER 4

Symbiosis

SYMBiOSIS was no doubt recognized by our ancient ancestors who, as they organized their world, saw relationships between plants and animals; this was not a recent phenomenon. In the modern world, we see a relationship between bees, flowers, and pollination, our gut bacteria and food digestion and the production of essential amino acids, and so on. This is only one "type" of symbiosis, that is, where two or more species cooperate for mutual benefit, again, like bees and flowers, where bees get simple carbohydrates (sugars), and the flowers get their pollen dispersed. While symbiosis of this type is the process of one species developing a relationship with another, entirely different species for mutual benefit, parasitism is where one organism gains a benefit while the host suffers poor health or dies. Examples include parasitic worms (e.g., roundworms, pinworms, whipworms, tapeworms), lice, and fleas. Using parasites as an example, the relationship might not have been mutual at first but adaptations take place (see Guerrero 1991); it might not remain mutual, as with an overgrowth of yeast in the human gut leading the vaginal yeast infections and so on. However, these relationships can progress into another type of symbiosis. This is where one species combines with another, often different species, resulting in a new species altogether (see Douglas 2010: 5-12; Bermudes and Back 1991: 73 for different symbiotic interactions).

Type One Symbiosis (Mutualism or Reciprocity)

Type One symbiosis involves two types of relationships. I have already mentioned one type, bees obtaining nectar from a flower while the flower benefits by attracting a transporter for dispersing pollen. Other examples include the Egyptian Plover bird that will fly into the mouth of a sunbathing crocodile and clean its teeth. The crocodile doesn't seem to mind, gets it teeth cleaned, and the Plover get a meal. Or the relationship between Tanzanian wart-hogs and mongooses. Mongooses travel in packs, and, when they encounter a warthog, they mossy on up, and eat the fleas and ticks, regular hitchhikers (parasites) on these fierce beasts. Perhaps more interesting is the relationship between ostriches and zebras. Ostriches have a poor sense of smell and hearing but great eyesight while Zebras, on the other side, have a great sense of hearing and smell. They apparently team up and protect each other from predators. One has to wonder how in the world did these relationships originate? In the above cases it would appear that threat displays and aggression is just "shut off," suggesting, perhaps, that some form of interspecies communication took place, early on, and this was somehow coded into a predisposition for mutual benefit—at least between the species involved. How would it be possible to develop these types of relationships randomly? There would seem to be some sort of "invitation." The other Type One symbiosis is where one species permanently lives on or within another species. This synergistic relationship may not have started out for mutual benefit, as mentioned, but develops into dependency wherein if one is removed the other "suffers." Let me give a real time example.

The Gut Microbiome

The foods you eat support different types of bacteria, archaea, viruses, and fungi. Microbiota are thrown out of balance by excessive amounts of sugar, salt, fat, red meat/animal flesh in general, Sweet and Low, antibiotics, alcohol, high fructose corn syrup, etc. Plant-based diets appear to be the best for supporting a balanced microbiome and they include:

Variety of whole foods (to assure sufficient vitamins, minerals, and other phytochemicals);

Fiber (you need 60-100 grams a day—most people get less than 20); Water—one ounce for every 2 pounds of body weight, more if working out and sweating.

A disrupted gut microbiome equals malnutrition and cellular toxicity and can lead to the following physical and psychological conditions (not all inclusive):

Type 2 diabetes;

Skin problems including acne, psoriasis, premature aging;

Chronic fatigue;

Cancer;

Psychological problems (depression, anger, etc.).

The medical community until recently has ignored the part nutrition and cellular toxicity have on overall heath. We can define health in many ways but most would agree that you need "proper" nutrition (you need protein, carbs, fats, vitamins, minerals, and other phytochemicals) and this is best understood by looking at what our Miocene and, more recently, ancient hunter/gatherer ancestors ate on a day-to-day and week-to-week basis. And what did they eat? Well, they were fruit and veggie eaters, they probably weren't fussy and tried to eat everything, but "everything" (all the culturally defined "food groups") is not available in the same place at the same time, thus nature forced our relatives to sequence foods. The USDA tells us to eat all the essential food groups at the same meal. Well, that's a problem. The USDA reflects cultural habits rather than how your body needs to get food for proper digestion. Our guts are a product of over 20 million years of eating mainly flowers, fruits, nuts, tubers, and veggies, with occasionally a little animal flesh. Even using stone tools (c. 3 MYA), for processing scavenged animals, meat eating was more than likely seasonal, late in the year when fruit or vegetable material was in short supply. Scavenging animals and placing one's self in competition with powerful predators and scavengers is a risky business—but you have to eat. Most of the year it was safer and easier to eat fruit, nuts, berries, and tubers. Any suggestion that we were slapping twelve-ounce hunks of animal flesh on the grill one million years ago is speculation at best; in fact, this is very recent behavior beginning at the earliest 300 KYA (see Wrangham's [2010] informative ideas regarding our use of fire). Until you can kill at a distance, our ancestors obtained meat from scavenged animals, and they usually got there last. Killing at a distance begins with the balanced throwing spear, which shows up around 300 KYA. Yes, I know, the Neanderthals would sneak up on a deer and stab it to death. This was risky business and accounts for most if not all the injuries noted in the bones, and probably a short life expectancy. Analysis of Neanderthal dental plaque reveal they included a great deal of vegetable material in their diet; they were not primarily meat eaters. Our gut microbiome, just like all aspects of human anatomy and physiology, has changed, evolved overtime. Hunter/gatherers, for example, because of their dietary behavior of consuming one food group at a time (nature does not provide a smorgasbord in a single place), had to forage in large territories (in most cases) to gain necessary calories. In the process they were not fussy and consumed many types of fruits, vegetable material, tubers, nuts, berries, and so on. Their gut microbiome probably differed from ours; we have limited choices of food, we consume many processed foods wherein the original nutritional factors are absent, we eat dense carbohydrates (bread, pasta, sugar) in high quantities—a diet foreign to our ancestors. Hunter/gatherers were probably less prone to disease and infection, and some suggest that understanding their gut microbiome might be helpful in disease prevention (see Weyrich 2015). The point is our guts are designed to maintain a balance of specific bacteria, archaea, viruses, and fungi that support proper food digestion. These hitchhikers get a meal and we use, for example, the amino acids and vitamins secreted during their digestion processes. Because of millions of years eating mainly fruits and veggies, a plant-based diet helps us to maintain a balance of gut organisms so that no one group of microbes has a reproductive advantage over another. The North American balanced diet, as recommended by the USDA, is a danger to your health.

Your digestive system has essentially six major parts: chewing apparatus, esophagus, stomach, duodenum, small intestine, and large intestine. The stomach is an *acid* environment and primarily a holding tank designed to "sterilize" fruit and other plant material through the use of HCl (hydrochloric acid). Bacteria and viruses are ubiquitous—they are everywhere including the fruits and veggie our ancestors stuffed in their mouths, chewed up, and swallowed. The primary use of HCl is killing bacteria. HCl is also important for binding intrinsic factor to B-12, denaturing (unfolding) proteins, converting pepsinogen to pepsin, and so on. The duodenum is designed to bring the pH level up to alkaline status before dumping into the small intestine (an *alkaline* environment) where most of the vegetable material (vegetable protein, carbohydrates, and fats) is digested, with nutrients passing through the intestinal wall into the bloodstream. The large intestine is another holding tank where moisture is removed and recirculated. After many hours, waste is excreted from the body. There is no problem, health is assured, until we combine the veggie diet with animal flesh. Keep in mind our scavenging ancestors. They ate animal flesh when no other food was available; and they ate it by itself. Animal flesh is digested in the stomach—as long as there are no dense carbohydrates to interfere with digestion. Anyone who eats according to the USDA, eating animal flesh at the same time as bread, potato, pasta, and other dense carbohydrates will die of malnutrition. Why? Because when you put 6-8 ounces of steak (chicken, fish, eggs) in your stomach with carbohydrates you can't get the pH down low enough (around 2) to denature the protein and convert pepsinogen to pepsin for breaking down the protein to amino acids; your body doesn't need protein—it needs amino acids. When you mix dense animal protein with carbs, you are likely to experience indigestion—this is why there are so many pharmaceutical drugs and over-the-counter antacids. Many people these days have bacteria in the stomach (it actually shouldn't be there)—*Helicobacter pylori*, which putrefies animal flesh (and ferments the carbs) making the flesh unfit for human consumption, if the pH can't get down to around 2. Buzzards, jackals, and crows can eat putrid flesh but dogs, cats, and humans cannot. The medical community claims that *H. pylori* causes ulcers; I challenge this. *H. pylori* is an opportunist. As long as there is a sturdy mucous membrane lining the stomach *H. pylori* cannot get near the stomach lining to cause an ulcer. But when the mucous membrane degrades in the presence of an over production of HCl, or dehydration because of alcohol abuse or inadequate water intake, the stomach lining becomes inflamed and the bacteria find a new home.

Psychological

Unbalanced gut microbiome leads to malnutrition and cellular toxicity. Additionally, cells do not function well if deprived of certain nutrients or when full of toxins. Similar to the medical community, nutrition is not often addressed by the clinical psychological community as a "cause" of emotional problems. Psychologists need good information (not as yet provided by the USDA) on diet and detox, and an understanding of the emotional issues that derive from poor nutrition. Trying to "talk" someone out of depression caused by a diet rich in sugar and other refined carbohydrates, deficient in B-vitamins, etc., will help pay the mortgage but it is not very useful to the patient. There are also genetic issues connected to depression, especially stress reactions and methylation issues as discussed in Chapter 3, that are not amenable to talk therapy. The next generation of clinical psychologists might find training in genetics and nutrition useful.

Food Manufacturers

Enough has been written about the dangers of consuming too much fat, sugar, and salt, with eggs and meat being unhealthy one day and then the next research study says something entirely different. And then we hear about the goodness of fruits and veggies along with an unimaginable variety of vitamin, mineral, and herbal supplements, and the plant-based diet. The reality is there is no solid information on the "perfect" dietary regimen, generalizations, yes, but scientifically based specifics, no. For example, how much vitamin C (ascorbic acid) does the individual need a day? Or, the B-vitamins, how much does the individual need? The USDA has put together the MDR, or minimal daily requirement for the masses, but you might need a maximal daily requirement if you are sick or injured, or because of coding sequence predispositions. There are differences between individuals and groups in terms of nutritional needs. Primates lost the ability to produce vitamin C, a good Lamarckian example of use it or lose it. Our primate ancestors of the Miocene were primarily fruit eaters, with berries, flowers, nuts, etc., rounding out the day's foraging. They were consuming a large amount of vitamin C, more than the 25 mg designated by the USDA as MDR, and they no longer needed to produce their own vitamin C. Dogs and cats manufacture vitamin C, they are not specifically fruit and veggie eaters, and when they are ill or injured, produce many times what they would need on a day-to-day basis. Vitamin C is important as an antioxidant, for proper immune functioning, strengthening blood vessels, and so on. Vitamin C is water soluble, and, consuming 500 mg of Ester C per day and as much as 2000 mg if suffering from a cold/flu, or recovering from surgery or injury, is more reasonable than 25 mg. You know when you have overdosed on vitamin C if you get diarrhea. I suspect, and hope, in the next few years the importance of nutrition will lead to a better understanding of human needs and health, and much of the packaged and fast foods sales could be targets of lawsuits. Gut bacteria and the dependency created leads us into Type Two symbiosis wherein the combining of species leads to an entirely new species.

Type Two Symbiosis: Eukaryotes and the Creation of New Species

Bacteria and archaea both contain mitochondria (chloroplasts); these are the powerhouses of cells that take in nutrients, converting them into energy—this is called respiration. Unprovable but reasonable is the "endosymbiotic hypothesis" suggesting that mitochondria (and chloroplasts) descended from bacteria that somehow survived endocytosis by a different species of prokaryote (archaea) or some other cell type and became incorporated into the cytoplasm. But that is not all. Bacteria and archaea do not have a nucleus, but the next stage of complexity, the eukaryotes, do. An eukaryote comprises a cell or cells in which the genetic material (DNA) in the form of chromosomes is contained within a distinct nucleus. Eukaryotes include all living organisms other than bacteria and archaea. The point: bacteria can invade other bacteria, and if they are not destroyed by the host's immune system, a new species can develop as new data is added to the informational system. The addition or development of the mitochondria and the nucleus occurred early on, c. 3.5 BYA for mitochondria (perhaps earlier if delivered to Earth on comets or meteors) and 1.7 BYA for the nucleus, perhaps when the immune systems of bacteria and archaea were immature or unprepared for new strains of bacteria that could defeat immune responses. However, I've often wondered if viruses and bacteria can read the genetic code of one another. Viruses must test cells for entrance by fusion with the membrane, common in viruses with an envelope. They can also stick to the cell and allow/trick the cell into engulfing the virus (endocytosis), or they inject their genetic material into the cell where it gets copied and produces more viruses. These various processes do not seem to be random but appear specific to certain cells and organisms. Returning to the mitochondria, we know it is bacterial in origin because the DNA forms a round shape rather than looking like an "X" in eukaryotes. Somehow this bacterium survived endocytosis by another species of prokaryote or some other cell type and became incorporated into the cytoplasm. One of the chief supporters of symbiogenesis or the combining of different bacteria or species is Lynn Margulis (2003: 145-161). In fact, Margulis suggests (2003: 6), "[i]ronically the popular evolutionist's view that organism originated by the accumulation of random mutations best describes the evolutionary process of bacteria. All the large, more familiar organisms originated by symbiont integration that led to permanent associations." Her first conclusion, that random mutations produced the first bacteria, as noted in earlier chapters, is highly improbable and not realistic. Her second conclusion, that is, all eukaryotes are the product of symbiogenesis early in the development of life on this planet is probably correct. Margulis further comments (2003: 56):

> Details abound that support the concept that all visible organisms, plants, animals, and fungi evolved by "body fusion." Fusion at the microscopic level led to genetic integration and formation of ever-more complex individuals. The thermodynamic drive toward more complex gradient-reducing systems finds expression in creating newer, more intricate forms of association between life forms, including symbiogenesis.

She goes even further in suggesting (2003: 7):

> However, the "phylogenetic" or "evolutionary" or "cladistic" concept of species is entirely wrongheaded, and its adoption interferes with understanding how species arise. The long-term symbiosis that led to species origin by symbiogenesis requires integration of at least two differently named organisms. No visible organism or group of organisms is descended "from a single common ancestor."

One of the mechanisms bringing different species together for possible symbiogenesis is stress, or situations where animals are drawn together (prototaxis; see Margulis 2003: 98) to, for example, avoid being eaten or perhaps cold or another environmental stressor/catastrophic event. Any suggestion, however, is just a guess. I want to digress for a moment and then come back to symbiogenesis as a possibility for animals just "showing up" at the beginning of the Cambrian c. 550 MYA. Margulis (2003: 124) outlines the "come back" phases of life after a catastrophic event.

> Specialization increases in ecosystems as they develop. When forest or river ecosystems grow back after a fire, flood, or other perturbation, recovery begins with fast-growing populations of organisms of the same kind. Such pioneers (certain bacteria, algae, grasses depending on the particular system) all behave in much the same way. Their populations grow rapidly to fill the available niche-space. But, then, inevitably, they face environmental limits that restrict further growth. Lack of space, water, phosphorus, or other nutrients blocks further population expansion. At this time more slowly growing species, which were unable to settle the original environments, join the fast-growers to make more efficient use of nutrients, energy sources, water, or other limited environmental variables. The slowest-growing and most stable communities of organisms at a given location, the so-called climax ecosystem, ultimately replaces the pioneers, sometimes after a long succession of intermediate stages. The climax groups of plants, animals, and microbes tend to have the greatest diversity of species, the most complex interactions, and the highest energy efficiencies relative to their predecessor communities.

The earliest life forms on planet Earth, anaerobic-type bacteria, show up c. 4-3.5 BYA, and may have come in from meteors and comets during the Late Great Bombardment, which lasted anywhere from 20 to 200 million years (4.1–3.9 BYA?). What we term "Snowball Earth," actually two cooling events, occurred between 2.4 BYA and 580 KYA (see Lane 2015: 24-29). The first occurred c. 2.4 BYA and lasted around 300 thousand years, ending with CO_2 out-gassing from volcanoes. By 1.7 BYA we encounter a new cell type, the eukaryote, the symbiogenesis of a bacterium and archaea. This Snowball Earth event could have been the trigger for their union. The next Snowball Earth event occurred c. 8-600 MYA, the end of which came, probably, with more CO_2 out-gassing from volcanoes. This also led to the beginning of the Cambrian c. 550 MYA and many animal species with few if any precedents in the pre-Cambrian. This second Snowball Earth may have been the stress factor leading to numerous symbiotic relationships solving the mystery as to the sudden appearance of radically new life forms that burst forth, like Athena, fully developed (and armored in many instances) in the Cambrian Period. Also recall genetic dormancy mentioned earlier. Again, scientists have known for some time that symbiosis occurred between bacteria and archaea, evidenced by the mitochondria in eukaryotes. Viruses have also combined with plants and animals (becoming endogenous or HERVs—human-endogenous-retro-transposons, or retrotransposons), which can lead to rapid speciation by entering new information into prokaryote or eukaryote cells. Although the origin of viruses is unknown, some suggest they are breakaway segments of DNA or RNA. Viruses have probably been there from the beginning, and that beginning of life might extend back 10 billion years or more. As Sole and Elena comment (2019: 178):

> Viruses are found infecting all forms of life and have probably been around since the first cells arose, or perhaps even before them. Tracing back the origin of viruses is a titanic, almost impossible, endeavor because they do not form fossils, and the only source of information are molecular phylogenies and comparative techniques that have been extensively used to compare the DNA or RNA genomes of today's viruses, and reconstruct backwards their evolutionary history, hopefully, to reach their origins.

Zimmer (2015: 53-54) comments:

The idea that a host's genes could have come from viruses is almost philosophical in its weirdness. We like to think of our genomes as our ultimate identity. The fact that bacteria have acquired much of their DNA from viruses raises baffling questions. Do they have a distinct identity of their own? Or are they just hybrid Frankensteins, their clear line of identity blurred away? At first, it was possible to cordon off this puzzle from our own existence, treating it purely as a question about microbes. The presence of viral genes was merely a fluke of "lower" life forms. But today we can no longer find such comfort. If we look inside our own genome, we now see viruses. Thousands of them.

Although origins are obscure, the study of genomes reveals that viruses and bacteria have played an important part in altering their host's response to environmental stressors by modifying or adding new information to the system. Are these relationships (information sharing) random or is there a consciousness here where thinking and choices are being made? The Neo-Darwinists would most likely have us believe this information sharing is random, a fluke of nature, but there are other possibilities. Viruses, for example, often attack very specific tissues, i.e., mucous membranes connected to the nose, throat, and lungs. The SARS virus (Severe Acute Raspatory Syndrome) can infect many cell types in several organs with immune cells and pulmonary epithelium as main sites of injury. The Coronavirus (SARS-CoV) is nothing new and is a member of the family Coronaviridae (see Cheng et al. 2007). The coronaviruses appear to represent a crossover from animals, especially exotic animals (possibly fox bats), brought into the Chinese workplace as the economy expanded and new businesses were established. This brought people and animals together in an intimate setting, prompting the viruses to alter coding and attack a new species, *Homo sapiens*. Because of the emerging and reemerging nature of this virus family over very short periods of time eliminates this as an exercise in randomness. The virus wants immortality, as do all life forms, and immortality is a futuristic adventure requiring active participation in that quest. Waiting around for the right mutation to occur is a death sentence. The coronavirus (COVID-19) attacking humans is not as deadly as SARS and this might suggest a random mutation rather than a purposeful alteration (see Behe 2007). However, becoming less virulent, with a longer incubation time, allows the virus to move quietly through a population—if the virus kills quickly it will invariably kill fewer people. A longer incubation period would allow the virus to affect many more people. This appears more of a strategy than a simple random mutation. Although less virulent than SARS (although this may turn out not to be the case), COVID-19 is more contagious as it can infect more people leading to a higher mortality rate (see Ricard and Hood 2020).

I bring up SARA and the recent coronavirus because viral attacks are not always positive. But one has to consider that a virus, like all life forms, wants immortality—culling the host may not be the best strategy in the long run. Many countries, including China, involve themselves in weaponizing bacteria and viruses, and one cannot remove the possibility that COVD-19 "escaped" from a lab somewhere near Wuhan. Unfortunately, political parties in the US will now make COVID-9 a political issue – instead of seeing this as a warning and preparing for the next wave.

Continuing with Type Two symbiosis, scientists discovered a peculiar gene in the human genome that encoded a protein made only by cells in the placenta. Labeled *syncytin*, cells producing this protein are located only where the placenta contacts the uterus, producing a layer of cells essential for transporting nutrients from the mother to the fetus. The more interesting part is that the protein, syncytin, necessary for fusing the two types of cells together is not from a human coding sequence—it comes from a virus (Ryan 2009:146-150). Moreover, this viral exchange regarding the placenta has happened more than once in the animal kingdom. As mentioned earlier, viruses are there from the beginning, using life forms to replicate themselves. Viruses and self-replicating cells represent a good example of the chicken and egg dilemma—which came first? The bacteria (or however we want to label the early self-replicating forms) needed viruses for acquiring novel information and it doesn't make much difference as to the function of the material gained from the virus. Once conjoined with the bacteria's DNA it would be new and perhaps useful—or deadly. As Ryan (2009) points out, viruses have been with us for hundreds of millions of years. They typically infect eggs or sperm, inserting their own DNA into ours. There are 100,000 known fragments of viruses in the human genome, making up over 8% of our DNA. As one would expect, chimps, gorillas and monkeys share the same coding sequence for syncytin and this sequence is very similar from one species to the next. More recently a second version of syncytin (syncytin 2) was found in humans and other primates. Syncytin functions not only in producing a placenta but aids in lowering blood pressure and restricting the immune system's response to foreign tissue in the system. Symbiosis is not a patch, as suggested by Leisola and Witt (2018: 207-211), helping to "save" Neo-Darwinism from extinction. In fact, it does the exact opposite; viruses usually attack specific tissues—that is not random and thus the Neo-Darwinists can't claim this as promoting their position. Regarding the placenta, the viral strategy may be to build a bridge for retroviral transmission from mother to offspring and from offspring to mother (see Haig 2012). You will have trouble selling me on the idea this behavior is random.

We are considering life forms and adaptation and survival. And what is trying to survive? Coded information, with the phenotypic expression—you and I, as a test. If the phenotype is a test then for sure the organism can think and decide or why have a test of the genetic coding in the first place? Cells seem to be smart enough to abandon old configurations for new and these new configurations we call species. To borrow a phrase, "life finds a way," and part of our problem is how we interpret our observations. I need to back up a bit. One of the earliest scientists to realize the importance of symbiosis, as mentioned, is Lynn Margulis (1999). Although she assigns a limited role to random mutations, she believes they are responsible for building bacteria. The next cell type, however, is a symbiont. She states, "Random DNA mutations, primarily destructive in their effects, account only for the beginnings. *The role of randomness has been exaggerated in the evolutionary saga"* (Margulis and Sagan 2002: xv). (emphasis mine)

By "beginnings" she suggests that bacteria and archaea (prokaryotes) were randomly "created," while eukaryotes are a product of a symbiotic relationship between bacteria and archaea. Her generalization is simple: Species arise out of symbiosis, not random mutations (2002:11). Margulis certainly narrows the field regarding cause and effect, relying on symbiosis to answer most questions regarding speciation. Randomness cannot explain the origin of bacteria. Again, I challenge the Neo-Darwinist to show the world, using materialistic science (math, physics, and chemistry), how a cell structure is built incrementally over a long period of time. What is the process? What are the initial phases of construction? What is the math and chemistry? There aren't many ways around this and my suggestion is that the cell formed in the wave function; there it could be assembled piece by piece because it doesn't have to function until the wave form is observed or measured. How did it form in the quantum state? The quantum state is not a bunch of subatomic particles chaotically moving around. If it was chaos with no structure, where does the structure come from once the wave form collapses? Chaos, again is just a word meaning "the unknown."

Returning to viruses and their contribution to the placental mammals, the virus added new information and we go from Theria (marsupial mammals) to Eutheria (mammals with a placenta), the development of a new species, almost overnight. I have difficulty envisioning a placenta being made or constructed incrementally – conception would be impossible, unless the old system stuck around until the placenta was fully functional. Margulis (1999) and Margulis and Sagan (2003) offer many examples of symbiotic relationships that lead to new species.

Viral Symbioses

A virus cannot duplicate itself; viruses either gave up the ability or never had the ability to self-reproduce. That they appear to show up at the same time as anaerobic bacteria on this planet (c. 3.5 BYA) suggests a connected developmental history—which may be out of this world. In any case, the suggested stages for viral symbiosis/symbiogenesis are as follows:

1) Virus enters host (often specific to certain tissues); these are called exogenous retrovirus. By retrovirus is meant that viruses use DNA to replicate RNA, while bacteria and eukaryotes use RNA to replicate DNA.

2) Host immune system comes into play ("Us" vs "Them"). Bacteria can be invaded, for example, by other bacteria (including archaea) and viruses. In order to develop an immune system, there has to be some sense of "self" and self-preservation. Immune systems, then, are very old and were probably pushed into existence by other bacteria or more likely viruses. There are several possibilities that occur when a virus enters a host:

   a) Host destroys virus (endocytosis);

   b) Host is "culled" but some individuals survive. The virus certainly wants to survive with the least energy cost to itself, and killing the host leads to the production of lots of clones, but their survival depends on finding a new host;

   c) The virus wants to survive and it would appear that decision making goes on for, in time, the virus can become endogenous/symbiotic with host by modifying, silencing, or losing its lytic enzymes (e.g., reverse transcriptase) possibly leading to new species. In short, it goes from a lytic (infecting cells with its own DNA) to lysogeny, or symbiogenesis.

In humans these endogenous viruses are called HERVs (Human Endogenous Retroviruses) and they are very important in biological evolution. An example of this "switch" from lytic to lysogenic state mentioned above is seen in Bacteriophage λ an enterobacteria phage. λ (lambda phage, coliphage λ) is a bacterial virus, or bacteriophage, that infects the bacterial species *Escherichia coli* (*E. coli*). The wild variety of this virus has a temperate life cycle that allows it to either live within the genome of its host through lysogeny or enter a lytic phase during which it kills and lyses the cell to produce offspring (see Lederberg 1950 and Lederberg 1953). It would appear that life, the code for building a functional, self-replicating cell, to slow down entropy, will use any resource available to accomplish that task. Life also seems to have an affinity for life or the attraction of one life form to another. Symbiosis, through the special relationships that develop, appears to be a major factor in the origin of species. I want to bring in a statement that comes from a very different field, cognitive psychology, which expresses a "mental" or "cognitive process" that interestingly aligns itself with symbiosis.

> Conscious agents can combine to form new conscious agents, and these new agents can again combine to form higher agents, ad infinitum. When two or more agents interact, each retains is individual agency, but together they also instantiate a new agent. The more each of the agents in an interaction can predict its experience from its actions, the more integrated is their joint dynamics and the more cohesive is the new agent that they instantiate. The decisions and actions of a higher-level agent can, in turn, influence the dynamics of the agents in its instantiation (Hoffman 2019: 190).

Nowhere in Hoffman's overall presentation does he mention symbiosis; his interest is in how we perceive the world and why nature selects for our perception of fitness (how to effectively and efficiently interact with our perceptions) rather than our perceptions being "real." In short, we see icons, and interface with the real world, rather than seeing the world as it is. I look at my cat and I see an "icon," some generalization I can interact with. When I turn away and look at something else, my cat could be just a bunch of glittering lights. The cat is really there even when I'm not looking, but I don't experience the cat as it "really" is; I agree with Hoffman – the human nervous system isn't designed to see or comprehend what "really" surrounds us.

Unfortunately, Hoffman refers to natural selection as the "force" that selects for fitness over truth. I may have misread Hoffman, but it seems his brand of "natural selection," or nature, is calling the shots, acting on our perceptual systems without input from the cell— although he suggests that "agents" make decisions. I'm not sure how his position—which is very interesting and intriguing, would fair once the idea that cells think and decide, and that "natural selection," of the Darwinian type, takes second place to the activities of the cell. (Also see Fodor and Piattelli-Palmarini 2011 for an extensive discussion of problems regarding natural selection. Nature selects by chance, not choice, and it seems improbable and unreasonable to think natural selection of the Darwinian type could select for anything specific, let alone fine-tune anything. In my opinion, the cell [endocellular selection] is making most of the selecting—it is responsible for its own survival, and this is not random.) Hoffman's presentation ends with some mathematics, an encouraging sign, something totally lacking in the Neo-Darwinian model.

Lateral Gene Transfer

Lateral gene transfer (LGT) fits better under symbiosis than epigenetics, but this is a minor issue. Lateral gene transfer, again, refers to the movement of coding sequences from one species or organism to another (see Ochman et al. 2000; Dunning Hotopp et al. 2007; and Gillings 2016). We especially note this since the advent of antibiotics. Although "random" mutations are suggested for bacterial resistance to antibiotics (Behe 1997), with the bacteria losing information in the process (Behe 2019), lateral gene transfer can lead to new species by adding new information to the system. Moreover, and as I hinted at in Chapter 3, it is unlikely that this process is random; some information exchange most likely occurs before the bacteria or virus makes contact and transfers the actual coding sequence. Why do I say this? Well, it is called an immune response; foreigners are not always welcome. Is it possible that bacteria can read the DNA code of other species and can either turn on or off an immune response? Or, perhaps the vector senses information that the bacteria contain that could be useful or at least adds other choices? This is purely speculative. But how do the bacteria circumvent the immune system so that information can be transferred? Bacteria and viruses must sense and circumvent or create docking sites on the sugar-coated cell walls, but perhaps there is more to this informational system. We know that plants communicate with one another through smell and other signals, as do animals (see Chamovitz 2012). In fact, our social communication systems might seem rudimentary compared to the informational systems wrapped around viruses, bacteria, and the life sustaining cells they desire to inhabit. Again, this is speculative, but "communication" between species has proved to be more complex than once thought.

Life's Default System

It appears to me that bacteria, archaea, and viruses are life's default system, and, if this the case, information for building these default systems was available at the beginning of the universe, in the wave form, before it was "observed" and collapsed into the building blocks of the universe we experience today. You don't need the instructions to build a velociraptor or a human being; you only need instructions for building a cell and all the systems for that cell to function and duplicate itself, and probably a virus. Epigenetics, symbiosis, and hybridization (Chapter 5) would, over time, bring together new information for developing new forms, especially a more complex form of life, eukaryotes.

Further, bacteria and viruses can tolerate extreme conditions (some are called extremophiles), including extreme heat, cold, and high salt and radioactive environments. And then there are bacteria that can survive conditions found in interstellar space by forming spores, or a form of the bacteria that goes dormant but carrying all the coding sequences as found in its active or metabolic state. Is it possible, as mentioned earlier, that the spore, existing within the wave form, can adjust its coding sequences? If this is the case, going dormant after a catastrophic event might be another key ingredient in the development of new species. As Margulis comments (2003: 41):

> Bacterial spores and carefully frozen embryos do not lose their intrinsic organization when frozen, and upon warming they are fully capable of the metabolic transformations that underlie self-maintenance. Hence, we can see that although life fully blooms only as a process of flowing matter and energy, it can remain dormant in the organization of living matter.

Conclusion

I think Margulis (2003: 12) sums it up the best:

> What is symbiogenesis and how is it related to symbiosis. First, what is symbiosis? Symbiosis is simply the living together of organisms that are different from one another. When originally defined by Heinrich Anton de Bary (1831-1888), symbiosis was the living together of "differently named organisms." Symbioses are long-term physical associations. Different types of organisms stick together and fuse to make a third kind of organism. The fusion is not random. Symbiotic relationships occur under specific environmental conditions. In some of these relationships, one partner in the symbiosis feeds off the other to its detriment and even death. Such exploitive associations are called "parasitic" or "pathogenic." They tend to be highly sensitive to environmental stress. The parasite that invariably and virulently kills its partner kills itself. With time and circumstance the nature of associations tends to change. The relationships that interest us most here are modulated coexistence between former predators, pathogens

and their hosts, their shelter and food sources. As members of two species respond over time to each other's presence, exploitative relationships may eventually become convivial to the point where neither organism exists without the other. Long-term stable symbiosis that leads to evolutionary change is called "symbiogenesis." These mergers, long-term biological fusions beginning as symbiosis, are the engine of species evolution. Symbiogenesis, then, is a possible end point of symbiosis. These interspecies interactions are not random and, *pushing aside the concept of randomness, decision making and planning for the future present themselves as more viable alternatives for understanding the origin of species.* (emphasis mine)

CHAPTER 5
Hybridization

HYBRIDIZATION is the combining of genetic materials typically between two closely related species, although the number of chromosomes can differ. Usually the species in question split from a parent group and eventually moved great distances from one another; inbreeding can occur in these separate groups, which can have deleterious effects. Over time—perhaps thousands of years, the groups rejoin and hybridize. Hybridization can also occur between different species. Let me say from the beginning that hybridization is not always positive. When positive, more robust hybrids occur, and they are called positive heterosis; when negative, negative heterosis (see Pfennig et al. 2016; Mesgaran et al. 2016; Todesco et al. 2016; Keller et al. 2010; Rudan et al. 2003). My wife and I manage a feral cat community and we see the negative results of inbreeding (brother and sister, father and daughter, etc.) after two or three generations. The cats develop immune problems exhibiting many allergies, seizures, herpes virus in the eyes (not contagious—transmitted genetically, and perhaps an example of a virus moving toward lysogeny), and polydactylism (usually with one or two extra toes on front feet) where extra digits are not functional yet the claws often grow into the pads of the foot. Negative effects of inbreeding for humans is well illustrated by Ashkenazi Jewish Heritage, the BRCA 1 and 2 coding sequences, and the risk of breast cancer (see "Medical Genetics of Jews," Wikipedia). Reich (2018) also discusses the caste systems in India which isolate groups, and, because of social/religious taboos, end up inbreeding. Species isolation, through bottlenecking (movement between two land masses which eventually closes off stranding groups geographically), catastrophic events that kill off large numbers of a particular clade, and so on, not only divide groups but alter the quality and quantity of genetic materials available to the survivors. Sometimes, this can be positive by eliminating negative, recessive alleles, but, on the other hand, adversely, by removing positive alleles—natural selection operates on chance, not choice. As time goes on the separate groups add new adaptive coding sequences and neutral or negative sequences. Once groups separate there is the problem of acquiring new alleles from fresh breeding partners. If there are no groups within, say, reasonable walking distance, no new material can enter the system, inbreeding can cause the accumulation of negative coding sequences, and extinction of the group is a real possibility. Remember, entropy is the loss of energy in a system and mutations of the Neo-Darwinian variety, and their accumulation, is a sign of inherent entropy—you can't defeat it, you can only slow it down. The claim of Neo-Darwinism is that random mutations lead toward negative entropy; this is not true. Humans and all animals and plants age, and the question is, Why? Well, it's simple. No system is perfect and just as cars and computers break down, so does the human cell. Over time efficiency decreases and this seems to be related to energy use. The more energy an animal uses, the shorter the lifespan—except for the human animal (see Johns 1999). Humans have a strange physiology and what Johns calls Life Energy Potential (LEP). We expend more energy per kilogram than any other animal and should live shorter lives. Humans have longer than what-would-seem usual Maximum Life Span (MLSP). When we consume calories, we produce oxidizing agents that destroy cells and likely effect our coding sequences. We produce fewer antioxidants in our bodies (endogenous antioxidants) than almost any other mammal. How can this Be? Our ancestors were fruit, tuber, berry, nut, and veggie eaters; these were a source of nutrition/calories and their pharmacy, their antioxidants. Meat eating, in any quantity, is fairly recent in human history. In short, we are addicted to plants, and the chemicals they contain, to inhibit cell damage, protect the immune system, strengthen blood vessels, clean out our livers, and so on to promote health. We need phytochemicals, the plant chemicals, and minerals from the plants and the earth itself (pica), for, after all, we are made from the dust of the universe. Individuals in a nutritionally deficient environment will suffer more negative mutations because of nutritional inadequacies. As I discussed in Chapter 3, individuals and groups can adapt overtime to nutritionally inferior diets as noted in the Dutch Hunger study. An interesting feature of the study is that we see common effects of malnutrition, not just on one or two individuals, but a generalization across the study. For modern humans our health has suffered, especially once we became sedentary agriculturists which, to begin, limited our variety of foods. Moreover, we have a biology not adapted to the consumption of dense, processed foods (white, processed carbohydrates), nor is it adapted to eating the "balanced" meal as suggested by the USDA. In the West, at least, we encounter cultural attitudes that do not recognize the importance of antioxidants, etc., in the diet. The point is that those groups living in a nutritionally dense environment are likely to be more fit than those in nutritionally deficient environments. Alterations to the coding sequences in those environments will differ from those in a nutritionally deficient environment. We cannot attribute this to random mutations. Moreover, the hybridizing of these extremes often produces offspring combining the best of both worlds. Why? Most deleterious mutations are usually recessive, but recessive alleles can be expressed with inbreeding. Hybridizing often eliminates these recessive alleles leading to more "fitness." How often and under what circumstances does such hybridization occur? As mentioned in Chapter 4, one mechanism that could bring distant species together for hybridization is stress. The term for this is "prototaxis" (see Margulis and Sagan 2002: 98) or situations where animals are drawn together, for example, to avoid being eaten or perhaps cold or another environmental stressor/catastrophic event. Again, this is just a guess.

Examples of Hybridization

The Wikipedia website ("List of Genetic Hybrids") outlines many hybrids, from a cross between insects (bees are a good example), a California tiger salamander and a Barred tiger salamander and other reptilian mixes, to Zonkeys (zebra/donkey hybrid), Liger (lion/tiger hybrid), Jaglion (jaguar/lion hybrid), and many others (see Arnold et al. 2015). Not all the hybrids would have happened in the wild, and, accept for the human species (which will apparently have sex with anything), sexual signaling would be one issue to mating. The correct colors, displays, sounds, and so on would be necessary for attracting a mate, and then there are the mechanics involved. For example, I've never seen or heard of a dog having sex with a domestic cat, a dangerous adventure even for another cat. The signals are all wrong, and mechanically, the dog doesn't have the right equipment. But humans have had sex with a wide variety of animals—horses, sheep, goats, bulls, chickens, and perhaps gorillas – such encounters, to my knowledge, have never produce a viable zygote, although I'm sure that won't stop adventurous souls from trying. A story appeared in the news some years ago (Khamsi 2007) regarding the origins of pubic lice (*Pediculus humanus*). Apparently, around 3 MYA one of our early ancestors, an early erectus-type, acquired these from an archaic gorilla. The author is kind enough to our ancient ancestors to suggest that they probably obtained them while sleeping in abandoned gorilla nests, and not from sex. The analogy here is sleeping in a sleazy motel and taking on board bed bugs. First of all, pubic lice do not leap; they are close body parasites. Second, would one of our ancient ancestors choose to sleep in the nest of another animal? It's possible, but that would seem to me a dangerous enterprise as gorillas normally don't forage far from home base. Third, those early erectus-types looked more chimp-like than human, and it's difficult to appreciate their sense of "beauty." Perhaps the gorilla buttocks display was appealing to them, or perhaps this was just a "boys' night out," it's difficult to say. Anyway, gaining pubic lice requires close contact with more than just one individual. I have more to say about human hybridizing shortly. The point is, not all hybridizing is productive in the sense of producing viable offspring. Moreover, this is not to say that the viable hybrids in all cases are more fit (see Faria and Sucena 2015:115)—mules (female horse and male donkey), for example, are stronger, live longer, are smarter, and have a better temperament than their parents. However, mules, like some other hybrids, are sterile. We usually attribute the sterility to differences in the number of chromosomes leading to the inability to produce sperm or eggs.

Hybrid Primates

We reserve the term species for individuals who can interbreed and have viable offspring. Using a cladistic model, we can assume that conspicuous derived traits (traits absent in the last common ancestor) might indicate different species. The problem is the word, "species." As discussed in Chapter 3, many phenotypic differences encountered among the various mammal groups are the product of epigenetic additions and subtractions of chemicals, insertions, duplications, and so on, and the differences may not be enough to hinder animals, especially those living within territorial boundaries, from hybridizing. These boundary territories remind me of border towns we experience in our own time. For example, Tijuana, Mexico and Tapachula, a city and municipality in the far southeast of the state of Chiapas in Mexico, near the Guatemalan border and the Pacific Ocean, are border towns. Lots of interesting transactions go on in these places, transactions that center on drugs, guns, and sex. These are "no man's lands," where the rules are vague, there is little law, and lots of stress, and under such conditions people can end up doing unusual things. My wife and conducted research at the Mayan site called Izapa (almost on the border with Guatemala) some years ago and we had to pass through Tapachula to get there; I have some interesting stories to tell. The point is, border areas are a major concern for hybridization. Arnold et al. (2015) provide an interesting review of what they call "divergence-with-gene-flow" among numerous primate groups. Arnold et al. (2015:280-281 comment:

> In the light of the findings by Roos et al. (2011), it is significant that a series of earlier studies by Karanth and his colleagues (Karanth 2008, 2010; Karanth et al. 2008) concluded that the Asian colobine clade showed evidence of ancient hybrid speciation leading to the origin of the golden leaf monkey (*Trachypithecus geei*) and capped leaf monkey (*Trachypithecus pileatus*). Karanth et al. (2008) summarized these results in the following manner: 'The phylogenetic position of the capped and golden leaf monkeys remains unresolved. It is clear from both nDNA and mtDNA data that these two species are closely related. However, the mtDNA... tree strongly suggests that they belong to the Indian clade (Semnopithecus), whereas the nuclear encoded lysozyme gene suggests that they may belong to the SE Asian clade (Trachypithecus). Interesting, these two species are distributed in an area that is sandwiched between the distribution of Semnopithecus and Trachypithecus.' It would thus seem likely that not only has ancient and contemporaneous genetic exchange led to admixed genomes, but that such admixture has also led to new colobine lineages.
>
> Arnold et al. (2015: 281-282) note a similar situation with the gorilla where border areas seem to be the geography for hybridization. For the chimpanzee, however, they present a slightly different situation, stating, "A number of authors have tested the hypothesis of no divergence-with-gene-flow between common chimpanzees and bonobos (i.e., *Pan troglodytes* and *P. paniscus*, respectively)—the species considered closest to the human lineage..." This complete separation, however, is inconsistent with hybridization among other primates. However, around 1.5-2.5 MYA the formation of the Congo River might have created an obstacle for hybridizing.

The Hybrid Human

The hybridizing of most interest to me is that of *Homo sapiens*. There is some agreement in the anthropological literature that, with caution, our kind (using stone tool manufacture as a reference point) begins with something like *Homo habilis* or *Homo rudolfensis* and extends through time to you, me, and the neighbor next door. Using the original definition of "species," that is, to interbreed and produce viable offspring, from *Homo habilis* right up through to us, we are one species, a species that has hybridized repeatedly, the latest most likely being with Neanderthals around 55 KYA (see Reich 2018 for the genomic connection between erectus, Neanderthals, Denisovans, and modern humans within the last 300 KYA; also see Arnold 2015 regarding primates in general). For primates at least, as long as there is no more than 2 million years separating groups, they can interbreed and produce viable offspring. As I will hypothesis in Chapter 7, Homo ancestry may, in fact, extend back into the mid-Miocene; the Miocene (23-6 MYA) is the epoch of the ape, the base of the family tree. We see the existence of brachiators and quadrupeds, why not bipeds? This should not come as a surprise as all the codes for bipedalism were available during the time of the dinosaurs (see Chapter 7).

Frequency of Hybridization

Genomics, as time goes on, will help reveal the frequency of hybridization but it only has to be recognized one time in existing animals to be added to the list of evolutionary processes. As we can see from above, and what we know of primates and specifically the human animal, hybridization occurred more than once.

Conclusion

Humans have been selectively breeding and hybridizing animals for thousands of years. The dog, for example, *Canis familiaris*, ancestor to the European Grey wolf (*Canis lupus*), was domesticated c. 14 KYA, but 32 KYA has likewise been suggested. The wolf is a small group animal, an opportunist, and intelligent. We aren't sure how the wolf "turned into" the dog we experience today, but, just like humans, one of the first things to change, in both cases, is the face, the cranium. Our human ancestors, and this would have been modern humans and not Neanderthals (as far as we know), somehow selected those animals that were more subject to dominance by a human master. Dogs come in all shapes, sizes, and dispositions, and have been used to hunt, guard, kill, as food, and as companions for those in need of a second sight. Beginning around 10-12 KYA in the Zagros Mountains, and extending northwest into Turkey, we see domestication of sheep, goats, and cattle, and those animals that were dangerous and could not be domesticated were eliminated from the landscape. We can also appreciate that animal territories bump into one another and it is in these geographies that hybridization takes place and because of the stress generated in these locations, it would not be unthinkable that dissimilar animals might make for strange bedfellows.

CHAPTER 6

The Quantum World, Consciousness, and Reality

THE quantum world, we are told, is a strange place. Perhaps the math is wrong, or maybe our initial premises lead us toward "strange." Or perhaps we aren't able to see the big picture, and what we experience, we experience as random and without purpose; random is always relative. Conceptually/theoretically the quantum world comprises waves of energy that, when observed, morph into particles, keeping in mind that waves and particles are of the same "fabric" that informs the universe (the wave form IS the universe!). It's just that particles are more "open" to manipulation, or why have a particle in the first place? The quantum world reminds me of the Gnostic myth, where the "Parent of the Entirety" for some reason emits energy (*barbelo*), which then becomes the spiritual universe or universe within its wave form(s). Sophia, who represents wisdom (similar to Eve in the Old Testament), admires the spiritual (wave form) universe but sees it as unless and thus endeavors to make use of this energy by creating ("observing") a form, a particle named Ialdabaoth, who is a manifestation of the tangible universe. This would be Yahweh in the Jewish narrative, God the Father in Christianity, and Hiliah (Allah) in Islam—all three are demons demanding absolute submission, as with all demons. (Our mythic hero Jesus does not fit the definition of a demon.) Ialdabaoth then uses some of His energy to create other particles, Adam and Eve, who then create Cain, Able, and Seth. So, once the original particle (Ialdabaoth) is "created" the wave form collapses into the tangible world we experience today.

Hindus have a similar idea but compared to a play, wherein the energy that informs all (wave form) is powerful, quick, and can move back and forth between the Green Room (sanctuary of the wave form) and the stage (the collapse of the wave form) so rapidly (from wave to particle) that the audience sees separate particles—not just one huge particle or swirling mass. From this the individual believes he or she is separate from everything else. This idea of wave and particle is also evidenced in the Hindu narrative of Vishnu, the sustainer god or wave form, who dreams the dream of the universe. He is accompanied by Lakshmi, who, by rubbing Vishnu's feet, conditions, organizes, and sets rules to Vishnu's dream. And then there is Brahma, seated on top of a lotus flower issuing from Vishnu's navel. He is the creator god and personification of Brahman, the energy that informs all. We can see Brahma as the wave form's self-awareness, for when Brahma opens his eyes he sees the universe as it morphs from wave form to particle. Paraphrasing the storyline, we find Vishnu asleep on the cosmic serpent (death and life's renewal), with Lakshmi, his consort, rubbing his feet (in Hinduism you cannot have male without female or shakti energy). Vishnu, thus, with help from without, dreams the dream of the universe. Suddenly, emerging from Vishnu's naval is a lotus, sitting on top of which is Brahma, and when Brahma opens his eyes, a universe comes into existence. After so many billions of years, Brahma closes his eyes and that universe fades away, only to return after a few billion years. One Brahma day and one Brahma night equal 8,640,000,000 years (see Campbell 1974: 142-144); this is very close to the original calculation Edwin Hubble came up with in the early part of the Twentieth Century. Newer calculations put the supposed age of the universe at 13.8 billion years.

Entanglement

Carroll states (2019: 91):

> Entanglement arises because there is only one wave function for the entire universe, not separate wave functions for each piece of it.

He continues (2019:93):

> A wave function is an assignment of a complex number, the amplitude, to each possible observable outcome, and the square of the amplitude equals the probability that we would observe that outcome were we to make that measurement. When we are talking about more than one particle, that means we assign an amplitude to every possible outcome of observing all the particles at once. If what we're observing is positions, for example, the wave function of the universe can be thought of as assigning an amplitude to every possible combination of positions for all the particles in the universe.

Apparently, when we observe particles, we isolate them from the wave form comprising the universe, "just as if they have their own wave functions" (Carroll 2019: 96). Physicists freely admit that they do not know what is going on in the one wave form that functions for the whole universe. We experience bits and pieces of it with our sensory organs, and because we can manipulate these particles, there has to be a correspondence between our senses and the wave form; this is the "reality" we experience (see Hoffman 2019). Now, we may disagree on the interpretations of our experiences on one level, but, on another level, there is a tangible nature to the particle world we can agree on, thus allowing us to manipulate our environment.

One problem with quantum mechanics is the word "consensus" among scientists who attended the Solvay conference (1927) and the presentation of the ERP Paper (Einstein-Podolsky-Rosen 1935). Consensus is a fancy word for "we don't know for sure." In today's politics we have a similar situation with global warming; remember—we are in an interglacial period and it is supposed to be warming up. In science, something either is or isn't. Consensus among the "scientists" connected to the Catholic Church was the sun revolved around the earth and those who refuted this were treated very badly. When science enters politics, the data cannot be trusted, and if you don't understand why this is so, I suggest you get out of science and study sociology and our small group nature. Again, scientists really don't know what's going on in the wave form, producing, instead, models and best-guesses as to the rules that support that "environment." Quoting Richard Feynman, "If you can't reproduce it, you don't understand it." I agree with Einstein; our models are incomplete.

What Carroll (2019) leads up to is his bias (as he freely admits) in favor of a multiverse. As he states (2019: 122):

> The Many-Worlds formulation of quantum mechanics removes once and for all any mystery about the measurement process and collapse of the wave function. We don't need special rules about making an observation: all that happens is that the wave function keeps chugging along in accordance with the Schrödinger equation. And there's nothing special about what constitutes "a measurement" or "an observer"—a measurement is any interaction that causes a quantum system to become entangled with the environment, creating decoherence and a branching into separate worlds, and an observer is any system that brings such an interaction about. Consciousness, in particular, has nothing to do with it. The "observer" could be an earthworm, a microscope, or a rock. There's not even anything special about macroscopic systems, other than the fact that they can't help but interact and become entangled with the environment. The price we pay for such powerful simple unification of quantum dynamics is a large number of separate worlds.

From my understanding of quantum mechanics and the multi-world hypothesis, the universe began as a wave form which, at some point, was "measured" or "observed" by "something," thus collapsing the wave form into the observable universe. And, because the wave form is the universe, measurement or observation at any "point" would affect the whole universe. "What" took this initial measurement; "what" made the original observation, and where did this "what" come from? Is the wave form "conscious" and could it measure "itself?" Carroll (2019: 151-175) goes to great lengths during the dialogue between Alice and her father to minimize "measurement" and "observer," thus avoiding a major problem with the multi-world concept, that is, what started the shift from wave to particle? We need an earthworm, microscope, or rock and, again, the wave form(s) would initially have to be "measured" or "observed" by something. Carroll can't just push this under the rug by claiming that the "observer" is nothing special. Perhaps it is not, but the "observer" (or the interaction with the environment) needs defining and the "observer" itself would need to be initially observed. Because we don't really know what is going on in the wave form, any suggestion is open for debate. I do not believe in the many-worlds hypothesis. Instead, the "big bang" resulted from the wave form of the universe collapsing when it was observed by some "non-special" thing producing the one universe we experience today. Looked at this way there has been only one major quantum event since the beginning of the universe 13.8 BYA. However, we can't escape the possibility of a "big crunch" in the distant future, followed by another "Big Bang," followed by another "Big Crunch," and so on, a pulsating universe similar to that referenced in the Hindu tradition. But – this is speculation and we can't be sure. After the first "observation" or "measurement" all subsequent quantum events stay within the known universe, the wave form. Under laboratory conditions scientists have duplicated quantum events with electrons, but these quickly morph back in the wave form of the universe. Once that initial observation occurred those many billions of years ago, the universe was locked in. There may be different universes or dimensions, but do they come out of quantum events related to our universe? I don't think so. Then again, perhaps the wave form is an intelligence foreign to us—maybe we will never know.

There are two concerns in play that prevent thinking out of the box. First, Carroll goes to some lengths to cancel any idea that intelligence of one type or another had anything to do with the wave form morphing to particle. This is a negation of the Intelligent

Design issue and the suggestion that some intelligence made the first observation, putting into effect all the rules of the universe—including the development of self-sustaining life forms. The other concern is the question, is there a purpose to life? Neo-Darwinists and atheists alike claim there is no purpose to life, it is all random, with morality and culture, all inventions. I will have lots more to say about this in the Chapter 8.

Rules, Reason, and the Many Worlds

Looking at our world, the plants, animals, rocks, and all other conglomerations of particles we experience, we appear to exist within a set of rules and regulations, "products" of the wave form. I realize that many scientists consider this perception of order an illusion created somehow by our sense organs pushing us again toward random being the rule. Certainly, our sense organs perceive the world differently than, say, a wasp or a bat, thus we don't have a total experience of what surrounds us. Our sense organs function in a manner suited to our need to gain calories from the environment and pass our coding sequences onto the next generation (see Hoffman 2019). Because we don't have the total experience does not mean all is illusion. If random rules the universe, there would be no consistency to anything. There is nothing random about why it rains, or why the sun comes up in the morning; these are predictable events. Cause is predictable because we understand some of the, what we call, "laws of nature" (see Farnsworth et al. 2017 for an interesting discussion of causation). I have difficulty imagining a universe run on "random" rather than "rule." For example, during the early phases of the development of our solar system material was coalescing according to rules (electro/magnetic energy evidenced by electrostatic clumping, electrostatic discharge, heat, gravity), with larger chunks of material colliding leading to larger chunks, and so on, and in every case the trajectory of these masses can be calculated and collisions can be predicted. Order out of chaos is beset with rules and regulations. As Krakauer (2017: 171) comments:

Eugene Wigner, in his influential paper, "Events, Laws of Nature, and Invariance Principles" (1994), maintains that it is the role of physics to account for those regularities that are called 'laws of nature.'

The elements of the behavior that are not specified by the laws of nature are called "initial conditions." It is by definition of initial conditions that they are arbitrary with respect to physical laws, but as I have suggested, complex phenomena accumulate an alarming quantity of initial conditions and they are far from functionally arbitrary.

The wave form is no different. From what we know there is order (dense in the middle, less dense moving away) and energy variables. My interests lie not in position or momentum but in the "activities," or communication and information coding within the wave form. We know about spin-up and spin-down, right and left and so on and obviously there are rules in operation. With this in mind, I cannot see the wave form being able to bring forth all manner of combinations of ingredients in other assumed universes or worlds stemming from quantum events beyond the one we are experiencing. All things may not be possible. Going back to the Hindu and Gnostic narratives, we find a striking similarity with modern concepts of the wave form and particle worlds. And these narratives are by all means not limited to Hinduism and Gnosticism (see Joseph 2017b for quantum insights from Hinduism and other religious traditions). In Genesis, for example, we see, at least in the first creation story, a cosmology similar to current day thinking. What did the ancients really know and we are just now rediscovering, or perhaps rejecting, because it suggests the opposite of what the true believers in Neo-Darwinism most fear? How did they come to these conclusions? Did space aliens bring this to our ancient ancestors? Perhaps it was because of soma, a mind-altering brew (see Rush 2013), conceivably a combination of ingredients that would immerse us in the wave form and all the possibilities of the universe.

Or, perhaps, unfettered with dogma, they intuitively knew—perhaps it was self-evident to them. We are talking about beliefs that are at least 4000 years in the past and likely go back much, much further.

Why is the many-worlds view promoted by Neo-Darwinists? The main reason is to overturn Behe's (1996) concept of irreducible complexity and the statistics showing that the probability of building complex biological structures, such as blood clotting, the citrus acid cycle, and the cell, is slim indeed if dealing with one universe. If the universes are infinite, then anything is possible (at least within this thinking), and there would be at least one universe where life arose randomly—that universe is the one we live in; randomness and a non-purposeful scenario are the name of the game. Remember, the many-worlds concept is an opinion and if it is wrong, then many assumptions surrounding the concept are most likely wrong. Recall Carroll's (2019:122) statement:

> And there's nothing special about what constitutes "a measurement" or " an observer"—a measurement is any interaction that causes a quantum system to become entangled with the environment, creating decoherence and a branching into separate worlds, and an observer is any system that brings such an interaction about. Consciousness, in particular, has nothing to do with it. The "observer" could be an earthworm, a microscope, or a rock.

I'm not satisfied with the statement, "a measurement is any interaction that causes a quantum system to become entangled with the environment." What kind of measurement, and where did this "measuring devise" come from? Without understanding the "what" and "how" of the measurement, what follows is mere guesswork, storytelling, a narrative that avoids other possibilities. Here is another possibility. The wave form that is the universe, has rules connected to its existence, and because there are rules they had to come from somewhere—we just can't sweep this under the rug as being unimportant. In other words, the wave form represents some ordered communication implying intelligence. When the term "intelligence" is used the mind immediately refers to the human agency, but that's not the type or category of intelligence that exists within the wave form. Our intelligence, our consciousness is a sub-category of that whole, and it had to come from somewhere; it is not accidental.

All life forms have some form of intelligence, consciousness, and decision-making ability, which I do not believe spontaneously arrived. It seems reasonable to suppose that it emanates directly from the wave form proper, as the wave form contains all the rules and regulation for building the tangible universe and the self-replicating life forms within it, sort of like a *totipotent* cell that has the potential to develop all other cells and a complete organism. To deny such rules is likewise to deny rules of chemistry, math, and physics. As to the origins of the universe, that is, its emergences from the wave form, that event, apparently, is nothing special and of little concern, although it is easy to postulate/imagine/invent the multiverse as measured by the human element. I find it curious that human intelligence can measure electrons and note the change from wave form to particle, but we can't apply the same reasoning (intelligent observer) to the initial manifestation of this universe from the wave form which is the universe. In the laboratory it is purposeful intelligence, but for the wave form, that is the universe, the attribution is insignificant.

The Quantum World

What goes on in the quantum world, the wave form that morphed into the tangible universe we experience today? No one really seems to know. Yes, we have mathematics that lead to curious conclusions about multiverses, entropy, "spooky action at a distance," and so on—I suppose anything is possible, especially if you can wrap some mathematical formula around a cherished idea. Let's consider this: The wave form, that is the universe, is a living organism, unlike our bias of what makes up a living, self-reproducing organism. A major similarity between us and "it" are rules and processes of construction, of turning energy into matter and matter into energy, on one level, and turning what we term "inorganic" matter into organic matter and life-forms on another. Regardless of how you conceptualize the wave form, it must contain all the rules of the universe from the start—the wave form is *itself* the rules and processes. I find it very difficult to agree that once the "Big Bang" occurred all the rules of the universe randomly came together after the fact. In order for there to be a "Big Bang" there had to be some rules, some cause. Think about it. You first have to create randomly all the rules of the universe and then you have to create randomly the universe and randomly create at least the initial self-replicating life forms. Again, all this is going on, level after level of complexity, randomly! This is way beyond anything reasonable. All the rules, in my opinion, were available in the wave form along with all the energy to create the universe, but where would all the energy come from for "building" the infinite universes splitting off each time there is a quantum event? I guess in these other universes you can get something out of nothing—and this is a possibility in quantum mechanics. Looked at from another direction, we could imagine the wave form as a memory machine and this brings time into the equation.

Memory and Cosmic Consciousness

As of this writing no one to my knowledge can measure the mind, the brain, yes, but not the mind. Hoffman comments (2019: 150):

> For a theory that proposes that brain activity causes conscious experiences, we want mathematical laws or principles that state precisely which brain activities cause the conscious experience of tasting basil, precisely why this activity does not cause the experience of, say, hearing a siren, and precisely why this activity must change to transform the experience from tasting basil to, say, tasting rosemary. These laws or principles must apply across species, or else explain precisely why different species require different laws. No such laws, indeed no plausible ideas, have ever been proposed.

We assume our mind, our thoughts and memories, are a by-product of chemical and electrical activity in the brain—no more, no less. But is that all there is to it? Have you ever wondered why you can walk without falling down, or catch a ball or frisbee while running across a lawn, or why you can throw a football or baseball with such accuracy? It is because your mind in conjunction with brain hardware knows calculus, trigonometry, geometry, and high levels of math. If your brain wasn't programmed with these mathematical concepts, you could not function. Many great minds, in the past, have been able to take this math out of the "brain state," imagine these concepts in the "mind state," and then inscribe them on clay tablets, stone walls, or paper, and teach them to others. These mathematical systems were available at the beginning of the universe or the universe as we know it would not exist. And contrary to Hoffman (2019: 16) cats can "do" calculus — evidenced by their ability to leap and catch bugs and mice. What they can't do is move the calculus from the brain to the mind state and teach it to another cat.

Thinking, decision making, and what some consider consciousness appears to occur at the quantum level. As Perkins and Swain (2009:153) encapsulate in their abstract:

> Stochasticity pervades life at the cellular level. Cells receive stochastic signals, perform detection and transduction with stochastic biochemistry, and grow and die in stochastic environments. Here we review progress in going from the molecular details to the information-processing strategies cells use in their decision-making. Such strategies are fundamentally influenced by stochasticity. We argue that the cellular decision-making can only be probabilistic and occurs at three levels. First, cells must infer from noisy signals the probable current and anticipated future state of their environment. Second, they must weigh the costs and benefits of each potential response, given that future. Third, cells must decide in the presence of other, potentially competitive, decision-makers. In this context,

we discuss cooperative responses where some individuals can appear to sacrifice for the common good. We believe that decision-making strategies will be conserved, with comparatively few strategies being implemented by different biochemical mechanisms in many organisms. Determining the strategy of a decision-making network provides a potentially powerful coarse-graining that links systems and evolutionary biology to understand biological design.

Stochastic means random and this contradicts any thinking or decision making at the cellular level if "[s]tochasticity pervades life at the cellular level." This shows how the concept of random mutations, as the major cause in the development of life and the universe, has become imbedded in our thinking as the model into which research is forced. Again, these scientists assume randomness only because they cannot see the bigger picture. This is not good science. Invoking stochastic or random is another way of saying, "We don't know, life can't possibly be purposeful, therefore random rules."

> ...Many of the early founders of quantum mechanics held the view that the participatory role of observation is fundamental and the underlying "stuff" of the cosmos is process rather than the construct of some constant, underlying substance... quantum theory does not say anything specific about the nature of consciousness – the whole issue is clouded by basic uncertainty over even how to define consciousness... In our view, it may well be that the subject-object dichotomy is false to begin with and that consciousness is primary to the cosmos, not just an epiphenomenon of the physical process in a nervous system (Kafatos et al. 2017: 7).

> The fact that the wave function of the system collapses upon observation suggests that we should ask whether it fundamentally represents interaction of consciousness with matter... (Kak et al. (2017: 16-17).

Interesting and important experiments conducted by Libet and his co-workers show that the subconscious first decides and is then shared with the conscious mind (Libet et al. 1983). The suggestion, when considering the writings of other researchers, is that the subconscious mind, for example, when you dream, opens at the quantum level. In a matter-alone conception of the universe, we cannot conceive of a reality that has no objects in it, and only pure consciousness. However, such a matter-alone conception goes against the very nature of a quantum universe as quantum theory contains observation through measurement. Yet without awareness, nothing can be perceived.

Having placed its trust in 'reality as given,' science overlooks the self-evident fact that nothing can be experienced without consciousess... If enlightenment consists of seeing beyond the veiling that accompanies a commonsense view of the universe, it too isn't some sort of obscure mysticism but recognition that self-awareness can know itself (Kak et.al. 2017: 28).

It is more elegant, self-consistent and far easier to accept as a working hypothesis that sentience exists as a potential at the source of creation, and the strongest evidence has already been put on the table: Everything to be observed in the universe implies consciousness. Some theorists try to rescue materialism by saying that information is encoded into all matter, but "information" is a mental concept, and without the concept, there's no information in anything... In everyday life, we get to experience the miracle of transformation that causes a three-dimensional world, completed by the fourth dimension of time, to manifest before our eyes. The great advantage of experience is that it isn't theoretical. Reality is never wrong, and all of us are embedded in reality, no matter what model we apply to explain it (Kak et al. 2017:29).

Classical physics, by omitting all reference to the mental realities, produces a logical disconnect between the physically described properties represented in that theory and the mental realities by which we come to know them (Stapp 2017: 33).

The point here is that in orthodox (von Neumann-Heisenberg) quantum mechanics the physical aspect is represented by the quantum state, and this state has the ontological character of potential—of objective tendencies for actual events to happen. As such, it is more mind-like than matter-like in character. It involves not only stored information about the past, but also objective tendencies pertaining to events that have not yet happened. It involves projections into the future, elements akin to imagined ideas of what might come to past (Stapp 2017: 35).

What Stapp is suggesting is that imbedded in this "mind-like in character" is the ability to think ahead or plan into the future. Stapp goes on to say (Stapp 2017:36):

Furthermore, a quantum state represents probabilities. Probabilities are not matter-like. They are mathematical connections that exist outside the actual realities to which they pertain. They involve mind-like computations and evaluations: weights assigned by a mental or mind-like process. The neurodynamic processes under-pinning subjective consciousness are evolutionary ancient, are based on fundamental bifurcation events in biogenesis, and originate in single-celled Protista before the emergence of multicelled animals and nervous systems... Excitable membranes are universal to eukaryote cells, as is the need to sense electrochemical and nutrient changes in their milieu. The sodium channel key to the axon potential, for example, arose in founding single-celled eukaryotes (King 2017: 43).

(I)f the Second Law of Thermodynamics is to be preserved, it implies the existence of cosmological memory... By consciousness we mean any macroscopic or microscopic system that operates through the use of both memory and choice. Cosmic consciousness means, any cosmological system that requires both memory and choice to operate it (Christensen 2017:71).

A possible basis for the emergence of subjective consciousness, which could also be pivotal in explaining the source of free-will, is thus that the excitable cell gained a fundamental form of anticipation of threats to survival. These cells also evolved the ability to perceive strategic opportunities, through anticipatory quantum non-locality induced by chaotic excitation of the cell membrane, in which the cell becomes both an emitter and absorber of its own excitations. Non-locality in space-time is a fundamental quantum property shared by all physical systems, including macroscopic systems with coherent resonance... Quantum phenomena abound in biological tissues. Entanglement has been observed in healthy tissues in quantum coherence MRI imaging and bird navigation has been suggested to use entangled electrons. Excitation in photo-synthetic antennae have also been shown to perform special quantum computing. Enzyme activation energy transition states and synaptic transmission also use quantum tunneling (King 2017: 94).

Orchestrated objective reduction (Orch OR), a philosophy related to mind biology, suggests consciousness originates at the quantum level inside neurons, rather than the conventional view that it is a product of connections between neurons. As Penrose and Hameroff comment in regards to "proof" of consciousness (2017: 258):

The best-known temporal correlate for consciousness is gamma synchrony EEG, 30 to 90 Hz, often referred to as coherent 40 Hz. One possible viewpoint might be to take this oscillation to represent a succession of 40 or so conscious moments per second (t=25 milliseconds). This would be reasonably consistent with neuroscience (gamma synchrony). With certain ideas expressed in philosophy (e.g. Whitehead 'occasions of experience'), and perhaps even with ancient Buddhist texts which portray consciousness as 'momentary collections of mental phenomena' or as 'distinct, unconnected and impermanent moments which perish as soon as they arise.'... These accounts, even including variations in frequency, could be considered to be consistent with Orch OR events in the gamma synchrony range. Accordingly, on this view, gamma synchrony, Buddhist 'moments of experience', Whitehead 'occasions of experience', and our proposed Orch OR events might be viewed as corresponding tolerably well with one another. Consciousness would appear to be part of the evolutionary equation regardless of whether the reference is to inorganic or organic issues.

There are rules in play here and regardless of their origin, they certainly aren't random; they have to be the product of that which we know nothing. It's okay not to know something for it is not knowing and the desire to find out that pushes science for answers.

The Cell as an Information/Communication Specialist

According to Behe (1996: 41), "[w]hat a mutation cannot do is change all the instructions in one step—say, to build a fax machine instead of a radio." I agree with this regarding random mutations caused by some environmental factor (e.g. UV radiation, toxic plant chemicals. etc.). From the above quotes, however, it seems reasonable that cells decide, have choices at the very least in the phenotypic expression of an organism. A cell needs feedback, it has to know how it functions, and when the cell alters a coding sequence, it makes all other necessary genetic changes. This is not the case with random mutations, and if they aren't corrected usually lead to death and/or ill health of the organism. Neo-Darwinists do not consider this internally orchestrated change to be part of the evolutionary scenario.

This brings up genetic drift: is genetic drift (often compared to linguistic drift) simply random as well? Linguistic alterations are not like genetic alterations, wherein the alteration of a sound unit alters the pronunciation of a word and perhaps its meaning—such changes don't lead to the death of the language as it changes only a single word. Languages change through time with the alterations in phonemes (sound units), the development of new vocabulary to fit new technology, purposeful changes in ritual process, and so on. But these alterations are not like mutations, although alterations of sound units and spelling can be random; such alterations do not "kill" or weaken the language. It was once thought that the frequent use of taboo words shows a moral weakening of a culture and thus the language, and there may be some truth to it. In any case, genetic drift might just as well be purposeful as the organism senses the environment and alters coding sequences accordingly—making all necessary changes. I have labeled this morphological momentum or purposeful fine-tuning of systems. If genetic drift were purely random, there would be no life forms on this planet. The cell, in large measure, has to be responsible for its future. According to Hawking and Mlodinow (2012: 31-32):

> Though we feel that we can choose what we do, our understanding of the molecular basis of biology shows that biological processes are governed by the laws of physics and chemistry and therefore are as determined as orbits of the planets... It is hard to imagine how free will can operate if our behavior is determined by physical laws, so it seems that we are no more than biological machines and that free will is just an illusion.

With all the research available on the subject of consciousness, thinking, and decision making at the cellular level, and the speculation at the quantum level, I find this statement incredible. How can Hawking be so sure that all our choices are predetermined because of the laws of physics and chemistry? Where did these laws come from in the first place? This is a problem with those seen in "authority;" they can say such things, these narratives get repeated over and over, become fact, and graduate students and professors teach these opinions to students as fact assuming they are projecting intelligence in the process. But there is another part to this, and that is the assumption made about certain rules available in the beginning, especially the laws of physics. Where did the laws of physics, or the molecular basis of biology originate? Grisogono (2017: 66) comments:

> The foundation layer of such a framework has to be the differences that existed in the prebiotic universe— inhomogeneous distributions of matter and energy, arising from early fluctuations and amplified by natural processes with feedback, such as gravitational accretion of matter into stars, and the violent events in their life cycles. By definition, there are no observers yet to make measurements and to transform that latent information into what Bateson [1972] or von Weiszacker [1980] would have acknowledged as information. But we can surely be confident that the laws of physics were operating.

According to the above statement, all the physical laws were in place at the beginning, but only those that fit the materialist's model of the universe, that is, they can explain it all through physics, mathematics, and chemistry. According to the Neo-Darwinists, however, such laws had to be randomly invented after the fact, and that which we don't understand or can't measure using the materialist's yardstick, is ignored. I call this "hound dog science;" if you can't eat it, smell it, or have sex with it, urinate on it." The question is, where did these laws of physics, mathematics, and chemistry come from in the first place? In my opinion, if the "laws of nature" were available "in the beginning" all other laws, laws omitted or of which we have little or no understanding, were available as well. Us not knowing what is going on in the wave form, just prior to the "Big Bang" or shortly afterwards, should not exclude the possibility that all laws, including those of physics, mathematics, and chemistry, were there at the beginning. This would especially be the case if we live in a pulsating universe, for what has happened has happened a billion times before

— these pulses would never happen if there weren't rules in back of them.

Defining Consciousness

I define consciousness (see Rush 1996) in a rather limited way, with a reference mainly to hominins of the Homo genus beginning around 2.5 MYA (Million Years Ago). At this time in prehistory we see an animal capable of participating in or being part of his or her environment, stepping back (apart) from that environment, mentally extracting ideas or possibilities from that environment, and, through analogous thinking, "pressing" them back in. The first evidence of this is the manufacture of stone tools, made from special materials not found everywhere, and shaped for a special purpose. This is time displacement and implies memory and planning for the future and language. The suggestion is these tools, especially the Acheulian hand axe, are symbolic of the teeth of predators that were eating them—if this is true, this is perhaps our first evidence of magical thinking, of capturing the energy of another animal, and making it our own (see Rush 1996: 54-55), in a similar way that a cell might capture energy/information from a virus and thus enhance its fitness. So, our type of consciousness and symbolizing is designed around modifying the environment to fit our physiological needs, rather than altering our cellular structures to fit environmental needs. The new genetics, however, promises to point in the other direction, that is, changing our physiology for space travel or defense against disease, and so on. Both cases, however, offer more immediate solutions to environmental assaults in the future.

We know what our type of consciousness can accomplish, but where did it come from? Is it a matter of mathematics, physics, and chemistry, or is there something missing from our very successful materialistic model? To define consciousness in a broader scale, we first have to understand what it is we (as human beings) are conscious of. According to Bodovitz (2017: 309), "One of the defining characteristics, if not the defining characteristic of consciousness is the experience of continuity. One thought or sensation appears to transition immediately into the next, but this is likely an illusion." Bodovitz (2017: 312) continues:

> If consciousness vectors and a sense of continuity are necessary for consciousness, then the corollary is that changes in cognition are necessary for awareness... Even though an image may be static, our eyes never are, even during fixation. Our eyes are in constant motion with tremors, drifts and microsaccades. If these fixational eye movements are eliminated, then visual perception fades to a homogenous field. It appears that we are in a continual process of sensing information, going internal or going into some quantum thinking or organizing, and then out again in milliseconds leading to awareness of the outside environment.

Consciousness, to me, is metaphoric of the physiological process, and awareness is the end product or what we experience as tangible or real. From another perspective, quantum physics and the concept of entanglement ("spooky action at a distance") suggest that all is possibly preordained. As Joseph (2017: 335) points out:

> As demonstrated by quantum physics and entanglement, the future may affect and even direct the past or the present. Consider again entanglement between photons. In delayed choice experiments, entanglement was demonstrated among photons even before there was a decision to make a choice regarding these photons, that is, before it was decided to do the measurement. Joseph suggests that the present can influence the past, and this accounts for premonitions and déjà vu phenomena.

Joseph (2017: 356-357) has more to say about wave functions and consciousness:

> ...The wave function is the particle spread out over space and describes all the various possible states of a particle. According

to quantum theory the probability of finding a particle in time or space is determined by the probability wave which obeys the Schrodinger equation. Everything is reduced to probabilities. Moreover, theses particle/waves and these probabilities are entangled. Reality is a manifestation of wave functions and alterations in patterns of activity within the quantum continuum which are entangled and perceived as discontinuous, and that includes the perception of time: past, present, future, and consciousness. The perception of a structural unit of information is not just perceived, but is inserted into the quantum state which causes the reduction of wave-packet and the collapse of the wave function. It is this collapse which describes shape, form, length, width, and future and past events and locations within space-time... Consciousness can also reflect upon and become conscious of being conscious, and in so doing, creates a collapse of the wave function which is experienced as a dissociated consciousness observing itself; conditions which are not uncommon during accelerated states of brain-mind activity typical of terror and other emotional extremes... [D]issociation... also occurs when dreaming and can be attributed to a collapse of the wave function; consciousness creating itself by dissociating itself from the quantum continuum.

Jansen comments (2017: 393):

All physical theories are necessarily based on an observer who is the only one to collect observations and thereafter to transform them into physical theories. Only when the observer is not considered, a "theory of everything" could give the impression of complete knowledge similar to the philosophical concept of eternalism of Parmenides. However, the observer is still considered as an important factor in the conception of physical theories, his uncertainty about the unverifiable extrapolation to the far past and the final future has also to be respected. There seems to be a semantic problem with the concept of time. Time perception could only be an illusion, if visual perceptions reduced to physical wave lengths is also considered an illusion. But then every reduction to more basic physical entities should be an illusion and the word losses its original sense. Misinterpretation after unclear sensory experience remains the essential meaning of illusion, whereas any reduction to lower level physical factors should not be considered an illusion.

Consciousness is many things. We can consider it as a stream of thinking made up of stops and starts, as something that materialistic science will uncover, or even an entity that pervades the universe infecting all living things with the ability to define right and wrong, good and bad. However we want to define it, consciousness comes with an imagination and we do not have a filter for compartmentalizing right as opposed to wrong imaginings—which brings us to the multiverse.

Unique-Universe vs Multiverse

The multiverse model was first articulated by Everett (1957), and amplified by DeWitt (see DeWitt and Graham 2015). Vaas states (2017: 428-429):

> The central tenant of the multiverse hypothesis (henceforth called M) is that our universe is just one among an ensemble, an infinitesimal part of an elaborate structure which consists of numerous universes, possibly an infinite number of universes. Thus, M has been conceptualized and described by a variety of terms including " many worlds", "parallel worlds," "parallel universes," alternative universes," alternative realities," "alternative dimensions," "quantum universes," and so on. Some of these universes could have a physics or chemistry completely unlike our own, and this has led to additional questions, such as: is our universe fine-tuned for life, and if so, why?

Vass (2017: 429-430) goes on the describe the "different notions" of the multiverse, as it has been described as everything in existence which means there is only one universe; the observable areas of the cosmos that we can experience along with everything else that has interacted with, bumped into, or shared information with the observable universe; universes separate from one another but which make up "the universe as a whole;" different histories for the wave form that exist in superposition; "completely disconnected systems of universes;" and so on. Unless clearly defined by the writer, reference and context is not clear to the reader. According to Vaas (2017: 432):

> Unique-universe accounts take our universe as the only one (or at least the only one ever relevant for cosmological explanations and theories). It might have had a predecessor or even infinitely many (before the big bang) and/or a succession or infinitely many (after a big crunch), but then the whole series can be taken as one single universe with spatiotemporal phase transitions. From the perspective of simplicity, parsimony and testability it is favorable to try to explain as much as possible with a unique-universe. A both straightforward and very ambitious approach is the searching for a fundamental theory with just one self-consistent solution that represents (or predicts) our universe. (Of course, one could always argue that there are other, strictly causally separated universes too, which do not even share a common generator or a metalaw; but they do not have any explanatory power at all and the claims for their existence cannot be motivated in a scientifically useful way, only perhaps by philosophical arguments.)

Do scientists get things wrong? Do they formulate models, test them, and then, several years later with new information, discard them for new? The answer to the first question is a resounding, yes. But are old models thrown away? With Neo-Darwinism, the answer is, no. Instead of caving to Behe's (1996) "irreducible complexity," they grab hold of any model, no matter how "far out" to maintain faith in their product. The multiverse or many worlds interpretation (MWI) is no exception. As Mensky states (2017: 4290), "Nowadays it becomes clear that there is no purely objective quantum theory. Objectiveness is relative." The multiverse hypothesis, like Neo-Darwinism in general, are concepttualized by those authoritative academics as "scientific" even when scientific method is ignored. Scientific method requires: observations and questions about those observations, we render a hypothesis, intelligent researchers designs an experiment to test the hypothesis, and the experiments produce "facts," and perhaps another hypothesis changing the original. No such testing, using the techniques of materialistic science, has ever occurred regarding Neo-Darwinism. Instead, we are presented anecdotal examples that are supposed to stand as evidence.

Regarding the multiverse, we have computer programs, designed by intelligent designers, programs that can prove planet Earth is flat or, in this case, that multiverses exist. This isn't science; this is computer nerds designing software using scientific algorithms. But Vass (2017:436) cautions:

> First, data are theory-dependent, not just empirical; they might simply be false or interpreted wrongly. Thus, theories should not be thrown away too quickly.... Second, theories are often ahead of data. Thus, theories should make testable predictions of course, and observers and experimenters should obtain data. Supersymmetry models, for instance, were theories in search of applications initially, and physicists (and even mathematicians) learn-ed a lot from them although there has still not been a single supersymmetric particle discovered (yet). So, testability, a falsifiable prediction, is important, but not necessarily at least at the initial stages of theory building. Thus, theories must be given a chance to develop, to be purged of errors, to become more complete. Theories need to be tested and data collected, and new theories should not be discarded too quickly.

I would not suggest we discard the multiverse; how one would prove the concept, just like my position on genetic dormancy and irreducibly complex systems being assembled in the wave form, remains to be seen. The multiverse model is relatively new, in the sense of scientists taking the concept seriously; it should be given a chance to prove its worth. The belief that random mutations lead to new species, on the other hand, has been around for over 100 years with only anecdotal examples offered as proof. According to Behe (2019: 104-106):

> Invoking speculative theories of cosmology, bioinformatician Eugene Koonin (2012) proposes that we live in an infinite multiverse where any physical event—no matter how unlikely—that is not absolutely forbidden by physical law will happen an infinite number of times. Since an origin of life—complete with all the genes needed for the subsequent unfolding of life as we know it—is not absolutely forbidden, then it has happened by chance repeatedly, endlessly in some universe or other. Since we find ourselves to be alive here, then we necessarily live in one of those universes where life haphazardly arose. Koonin is quite serious and sober about his proposal. To show his good faith, he calculates the probability of life arising in a volume the size of our own observable universe and comes to a generous value of 1 in $10^{1,018}$. In other words, he agrees that the odds of life arising even in a universe with life-friendly laws like ours are beyond horrendously bad, well past vanishingly small. Yet, since he takes the multiverse to be infinite, the odds don't matter. I strongly criticized the infinite multiverse hypothesis in the final chapter of, *The Edge of Evolution*, pointing out its poisonous implications for science or, for that matter, any kind of knowledge about external reality. It leaves us no better off than thinking we're just a brain in a vat. What's more, it's contradicted by the apparent lushness of life, which seems to contain many more sophisticated systems than necessary to produce conscious observers. Here I'll bypass discussing the idea in depth. Instead, I will just point out that no explanation is offered for any functional aspect of life, everything of importance is simply posited as existing from the

beginning, the result of one humongous stroke of luck. There is no accounting for properties of systems, no reasoning from patterns, no appeal to processes we see in operation today except to say that, if it weren't this way, we wouldn't be here to make the observation. The entire account truly boils down to the mocking image conjured by physicists and Darwin skeptic Fred Hoyle—of a tornado that passes through a junkyard and assembles a jet plane—except that, to make it "reasonable," Koonin postulates an infinite number of universe-sized tornados and junkyards. Once you start invoking infinite multiverses to account for elegant biological machinery, it's hard to stop. Koonin uses it not only for the origin of life, but for the type of irreducibly complex biochemical system I discussed in *Darwin's Black Box*. The neutral processes (such as we've just discussed) that Koonin calls on to explain the rise of eukaryotes are incompatible with the evolution of complex molecular machines: general eukaryote features require neutral drift; functional eukaryotic machinery needs strong selection. So, since no actual explanation for them exists, the rationale seems to go, let's all just agree to say they are the result of our good luck in living in the right universe. Multiverse theory helps not a whit at accounting for life, because it simply posits cosmological unknowns to "explain" biological unknowns.

The larger issue is the fact that regardless if there are ten or an infinite number of universes, they all have to operate according to rules or laws. I've heard the statement, "chemicals have a natural affinity toward one another," with the overall suggestion that the building of carbon based (organic) chemicals then lead to life forms through randomness and chance. If this is the case, why are there chemicals in the first place? Why does carbon, for example, have 2 electrons in the first orbit, second orbit, and the third ($1s^2 2s^2 2p^2$)? Why does calcium have the following configuration: $1s^2\ 2s^2\ 2p^6\ 3s^2\ 3p^6\ 4s^2$? Obviously, there's a set of rules in play, and this would be the same in all universes; a universe without founding rules, in my opinion, does not exist—it can't exit. And where there are rules one should be skeptical of random activity producing much of anything, especially life forms.

Thinking and the Wave Form

Because many "scientists," for example Richard Dawkins, state they don't exactly know what is "going on" in the wave form, there is an assumption that whatever it is, it's random. Because the wave form is the universe, it is anything but a random soup of subatomic particles. Again, if the wave form is the universe, it isn't random. If this is the case, and I believe it is, then thinking, choices, and generic rules are in play—and this would have to apply to other universes as well. And, in my opinion, the generic rules, in our universe at least, include the ability to "think" and make choices at the cellular level. I think it unreasonable to believe that life forms are purely at the mercy of the environment without any say in their evolution. Dobyns (2017: 502) states:

> The quantum measurement problem remains mysterious, with multiple interpretations of this formalism making different assumptions about its nature. The various interpretations also make very different assumptions about the role of consciousness in this transition from super-imposed probabilities to definite observed events. Some regard it as completely irrelevant, some regard consciousness as essential to complete the transition, and at least one speculative model identifies this transition as the key physical substrate of consciousness itself.

According to Martin et al. (2017: 404):

> The projection of our subjectivity in the environment in which we live (synchronicity phenomena of type II), in agreement with quantum mechanics, refutes the local hypothesis ("each individual is in his parcel of space-time") as well as the realistic

hypothesis ("the object has a reality well defined independent of the subject who observes it"). As an end let us mention a quantum effect that can have important consequences in mental phenomena, for ex-ample for awareness (for the emergence of consciousness). It is the Bose-Einstein condensation, in which each particle loses its individuality in favor of a collective, global behavior.

According to Mensky (2017; 488):

...[T]urning off consciousness (in sleeping, trance or meditation) opens access to all classical alternatives put together, without separation between them. Of course, the access is realized then not in the form of visual, acoustic or other conscious images or thoughts. Nothing at all can be said about the form of this access. However, if we accept EEC (Extended Everett's Concept), then we may definitely conclude that the access is possible in the unconscious mind.

In other words, when asleep, in trance, daydreaming, meditation, and perhaps at death, we fall back into the wave form and have access to all the possibilities represented by the many worlds, if there are such things. "Thus, Extended Everett's Concept (EEC) leads to the conclusion that unconscious state of mind allows one to take information 'from other alternatives' that reveals itself as unexpected insights, or direct vision of truth" (Mensky 2017: 489). Gao states (2017:509-510):

It is shown that a conscious being can distinguish definite perceptions and their quantum superpositions, while a physical measuring system without consciousness cannot distinguish such nonorthogonal quantum states. This result may have some important implications for quantum theory and the science of consciousness. In particular, it implies that consciousness is not emergent but a fundamental feature of the universe... There are two main viewpoints claiming that quantum measurement and consciousness are intimately connected. The first one holds that the consciousness of an observer causes the collapse of the wave function and helps to complete the quantum measurement or quantum-to-classical transition in general (von Neumann 1932/1955; London and Bauer 1939; Wigner 1967; Stapp 2007). This view is understandable. Though what physics commonly studies are insensible objects, the consciousness of an observer must take part in the last phase of measurement. The observer is introspectively aware of this perception of the measurement results, and consciousness is used to end the infinite chains of measurement here. The second view holds that consciousness arises from objective wave function collapse (Penrose 1989, 1994; Hammeroff and Penrose 1996; Hammeroff 1998, 2007)... Though these two views are obviously contrary, they both insist that a conscious perception is always definite and classical, and there are no quantum superpositions of definite conscious perceptions... Dynamical collapse theories (Ghirardi 2008), many-worlds theory (Everett 1957) and de Broglie-Bohm theory (Bohm 1952) are the main alternatives to a complete quantum theory. The latter two replace the collapse postulate with some new structures, such as branching worlds or Bohmian trajectories, while the former integrate the collapse postulate with the normal Schrodinger evolution into a unified dynamic. It has been recently shown that the dynamical collapse theories are probably in the right direction by admitting wave function collapse.

According to King (1996: 538):

...the constant interaction between information coming from the past and information coming from the future would place life in front of bifurcations. This constant antagonism between past and future would force life into a state of free will and consciousness. Consequently, consciousness would be a property of all living structures: each cell and biological process would be forced to choose between information coming from the past and information coming from the future... This constant state of choice would be common to all levels of life and would give form to chaotic behavior on which conscious brain would feed.

Anticipatory pre-stimuli reactions seem to be incompatible with the Copenhagen Interpretation, since Schrodinger's wave equation treats time in an essentially classical way and rejects the possibility of pre-stimuli reactions (effects before causes). Dick Bierman tried to overcome this limit of the Copenhagen Interpretation with his CIRTS model (Consciousness Induced Restoration of Time Symmetry), presented to the PA 2008 conference (Bierman 2008). This model states that almost all formalisms in physics are time-symmetric. Nevertheless, the Copenhagen interpretation of quantum mechanics, which postulates the collapse of the wave function, introduces a break of time symmetry at the point of collapse. The assumption of CIRTS is that the brain, when it is sustained by consciousness, is such a special system that it partially restores time-symmetry and therefore allows advanced waves to occur. The time symmetry restoring condition is not the brain per se but the brain sustained by consciousness. The restoration of time symmetry is suggested to be proportional to the brain volume involved in consciousness. CIRTS considers consciousness to be a prerequisite of reality with special properties which restore time-symmetry. However, in CIRTS the rationale behind consciousness is missing and its special properties seem to arise from nothing (Vannini and Di Corpo 2017: 538).

Vannini and di Corpo (2017: 540) go on to say:

In the CIRTS model consciousness is a prerequisite of reality. In the syntropy model the feeling of life is a consequence of the cohesive and unitary properties of advanced waves. According to the syntropy model, any form of life has a feeling of life. Consequently, we would have a feeling of life also when no brain activity is observed. This would explain why all forms of life, even the most simple ones, show anticipatory reactions (Rosen 1985) and why, for example, patients during surgery in a state of anesthetic-induced unconsciousness ten to defend themselves, and subjects with no brain activity react and defend themselves when their organs are removed for transplant. According to the syntropy model, the feeling of life does not reside in the brain; however, the brain provides memory which allows us to remember and reason regarding our conscious experience.

Joseph (2017: 547) comments:

The act of knowing, of observing, or measuring, that is, interacting with the environment in any way, creates an entangled state and a knot in the quantum continuum described as a "collapse of the wave function;" a knot of energy that is a kind of blemish in the continuum of the quantum field. The quantum knot bunches up at the point of observation, at the assigned value of measurement and can be entangled. Consciousness perceives a blemish in the continuum, but which is still part of the continuum, even though it is perceived as distinct. The universe exists, because there is consciousness of the universe. This also means; consciousness must have come first, creating a quantum jump from singularity to duality. The universe may have become conscious of itself.

Conclusion

As the reader can appreciate, quantum consciousness and the whole system of quantum mechanics is a complex issue with lots of opinions whether or not consciousness is part of the universe or even if it is necessary. The position one takes is probably more politically dictated then out of a sense of scientific conviction. Consciousness and purpose, for example, are not part of the Neo-Darwinist model. None of the models, however, lead us to the origins of life, or how it came to be. My opinion is that the instructions for life, i.e., how to create organic molecules, how to create the nucleotides (base pairs), how to create RNA, DNA, and management proteins/ enzymes, and how to construct a cell, and all the machinery involved, are embedded in the wave form. Life or the possibility for life has to have existed as a generic informational system right from the beginning; this is opinion, not fact. Even in a multiverse narrative, there have to be rules and order right at the onset.

CHAPTER 7
Hominin Evolution

THE idea that humans descended from apes is often attributed to Carlos Linnaeus (1701-1778) who promoted binomial nomenclature, originally constructed by Gaspard (1560-1624) and Johann (1541-1613) Bauhin, for cataloging plants and animals. Classification involves cataloging specific/shared anatomical traits. As we saw in Chapter 3, using shared or dissimilar anatomical (phenotypic) traits, for example, the Woolly Mammoth and the Mastodon, and the Nautili of the Great Barrier Reef, are perhaps not the best method for "species" identification. The ape to human connection was first identified by Thomas Huxley (1825-1894) who, in *Evidence of Man's Place in Nature* (1863) most clearly made the connection (Huxley 2012). Darwin didn't get around to expressing the connection until *The Descent of Man* (2012, orig. 1871). And except in the Orangutan's case, native to the islands of Borneo and Sumatra, and Gibbons, who occupy southern China and northeastern India and the islands of Sumatra, Borneo, and Java, chimps and gorillas live in Africa. Because chimps and gorillas appeared more similar to us, Africa was designated as the place of human origins. After verbalizing this connection for so many years the concept of an African homeland has become fixed, although some anthropologists and geneticists (see Bae et al. 2017) still consider Asia a viable option, especially in the evolution of modern humans. The question remains: did we evolve from apes or is this an opinion stated so often that it has become "fact?" Recent genomic studies show it is much more complicated, and attempting to place modern human origins in one geography, either Asia or Africa, is not realistic.

Us humans, in fact, evolved in many geographies. And we are still evolving; hybridization is occurring at a rapid pace.

Hominins

The term hominin (tribe, hominini) refers to those bipeds most directly related to modern humans. At this point in history this would include *Home erectus*, *Homo neanderthalensis*, *Denisovans*, and some ghost lineages discovered during genomic research of which we know nothing. Because of the confusion surrounding *Homo erectus*, that is, the different species names, and whether these various species designations are valid, I lumped all together into *Homo erectus*-types (Rush 1996). The only *Homo erectus*-type discovered to date that, in my opinion, does not fit is "*Homo*" *ergaster*; I give reasons for this below. From *Homo habilis* of c. 2.5 MYA to *Homo Heidelbergensis* of c. 800 KYA, they are all one species, so obviously we carry much of the genetic material as did Neanderthals and Denisovans and could interbreed/hybridize. This hybridization, one would theorize, led to the development of at least three phenotypically different groups— modern humans, Neanderthals, and Denisovans. In short, what we see through recent genomic studies is phenotypic diversity in the same manner as we did with erectus-types. Speculation, and it's only a guess, is that these three groups emerged from an erectus-type (Rhodesian Man, also known as "Goliath") after 800 KYA, with phenotypic differences occurring rapidly in response to environmental stress factors. Keep in mind that erectus-types were in Asia long before modern humans, Neanderthals, and Denisovans – over a million years before (see Zhu et al. 2018). The Chinese erectus-types adapted to an environment much different from Africa, with different foods, pathogens, predators, and weather patterns and likely added new immune factors (HLA human leukocyte antigens).

Our Molecular and Phenotypic Connections to Apes

Apes "show up" during the Miocene (23-6 MYA) with *Proconsul africanus* (c. 23-14 MYA) standing as an example of a creature, unlike a monkey, that exhibited specific traits not seen before in the paleontological record—lack of a tail is one of them. The claim is that an earlier specimen, *Aegyptopithecus Zeuxis*, is an intermediate or transitional form, part monkey and part ape, and the precursor to Proconsul. Aegyptopithecus, showing up c. 38 MYA in the early Oligocene has some ape-like traits, especially the teeth, but unlikely a close ancestor to Proconsul. Miocene apes are called "dental apes" because, mostly, all we have are teeth. Apes have a characteristic cusp pattern called "Y-5," wherein the lophs, ridges of enamel connecting the cusps, flow to the center of the tooth creating a Y-5 pattern (5 is the standard number of cusps found on ape molars although sometimes there can be 6 or 7). Pig molars, like human molars, have the Y-5 pattern, and sometimes are confused by medical examiners as human. When you encounter a conservative trait peculiar to one group of animals, it is a good bet that they are related—especially the teeth and other conservative traits. Thus, Aegyptopithecus is *somehow* related to Proconsul, but the relationship would be analogous to the genus Homo and our distant relationship to Lucy (*Australopithecus afarensis*) because both have ape teeth (gorillas and chimps have ape teeth as well). Teeth are conservative, they change little over long periods of time and they last longer than any other part of the skeleton post mortem. Genomic research conducted over the past 20 years shows the connection we share with all apes. So, there is common descent, the one thing I believe Darwin got right; all the apes share a common ancestor that lived c. 25 MYA, at the end of the Oligocene and the beginning of the Miocene. We don't have the slightest idea as to the identity of this ancestor. To complete this part of the current narrative, Gibbons diverged from the common ancestor c. 17-20 MYA, Orangutan 14-18 MYA; gorilla c. 13 MYA; chimp 8 and 6.3 MYA (nDNA and mtDNA suggest a divergence 8 MYA and then a hybridizing c. 6.3 MYA). How was this determined? I will discuss dating methods later in this chapter. Let me say, however, molecular dating methods, including Potassium/Argon, U-238, and even Carbon 14 dating are problematical for several reasons.

How DNA Accumulates Changes or the Molecular Clock Model

We base molecular clocks on assumptions about biological processes considered the source of all heritable variation: "mutation" and recombination or "crossing-over." Crossing over is the exchange of genetic material between non-sister chromatids of homologous chromosomes during meiosis, which results in new allelic combinations in the daughter cells. Much of the current literature still refers to mutations and recombination as random and non-purposeful, and that DNA is the executor or "decision maker." The cell is the executor, and when we throw into the mix that cells think and make decisions, it become difficult to know if alterations to the base pairs are purposeful or random. Coding errors that accumulate over time (genetic drift) are thought to be random, and I'm sure some are, but not all. In a single human genome, there are around 70 nucleotide changes per generation, or so we are told (see Kong et al. 2012). This is small when you consider the human genome is made up of 6 billion nucleotide letters. Over many generations, however, these changes lead to a great deal of variation and new species. Scientists can use these alterations to suggest when different groups split and went their separate ways. By comparing the coding sequences of two different species, say chimp and human, and knowing the rate of these changes (70 nucleotide changes per generation), researchers can calculate the time needed to accrue these changes, thus showing the time of divergence from a common ancestor. As you can imagine there are some problems with this. Alterations in nucleotides per generation can differ between species indicating that the mathematical outcome is a "best guess." Changes in environmental conditions can affect accumulation. In short, 70 nucleotides per generation is only a prediction and the dates giving above could be several million years off the mark.

Neo-Darwinism and Hominin Evolution

Hominins are us, or Homo genus bipeds; hominids, on the other side, comprise all bipeds, including the Australopithecines and Paranthropus varieties. At one time the robust boisei, robustus, and aethiopicus were placed in the same genus as the Australopithecines. For the Paranthropus varieties the differences in such conservative traits, i.e., the teeth and the systems that support them, show a very different food quest than the Australopithecines, such as Lucy (*Australopithecus afarensis*). This also indicates a very long period of separation and that the large, Paranthropus varieties should never have been placed in the same genus as Lucy. These robust varieties are very similar to one another with very large cheek teeth and accompanying large mandible and maxilla, large and buttressed zygomatic arch, the upper border of which gives attachment to the temporal fascia and the lower border and medial surface give origin to the masseter, and a sagittal crest for the attachment of the large temporalis muscle. The robusticity of these muscles and structures shows these hominoids could eat a large variety of foods including those that were tough and fibrous; we can only guess what they preferred.

We date the Paranthropus varieties to around 2.6 and 0.6 MYA, they would be contemporary with the Australopithecines, and therefore could not have evolved from them, although they may have a common ancestor. To look at this in a slightly different way, we consider *Sivapithecus indicus* to be an ancestor of the modern-day Orangutan. This is based on facial characteristics—eye orbits, nasal aperture, zygomatic bone, position of incisors, and so on. Paranthropus sports characteristics similar to modern gorillas especially in, for example, the sagittal crest, brow ridges, and nasal aperture and they may share a common ancestor. Paranthropus, however, lacks a fighting tooth, and, as a biped, had no use for the extensive nuchal crest possessed by the gorilla. There are suggestions (see Stokstad 2000 and Richmond at al. 2009) that the common ancestor to humans and chimps might have been a knuckle walker. But we could interpret this another way—chimps might have been, originally, bipedal and after 6 MYA reverted to knuckle walking and quadrupedalism; the find in 2000, *Orrorin tugenensis*, at least suggests bipedalism was available at that time. It is also very curious that we have yet, except for some chimp teeth dated to about 500 KYA, to find quadrupedal ape fossils alongside Hominid finds. However, *Ardipithecus ramidus*, and contrary to White and Lovejoy's reconstruction as a biped, was probably a quadruped and might be an archaic chimp (see description below).

Perhaps our opinions regarding the development of bipedalism are wrong. For example, Deloison (2004: 45-46) comments:

> I suggest a common ancestor for hominoids: proto-hominoid. Protohominoid had a small slender body with thin bones (not found now because they fossilized poorly and because we have not looked for them). It walked erect and was bipedal and plantigrade. Its anatomy showed little differentiation: a round head, almost equal limbs, pentadactylism and battle-shaped hands and feet. It must have lived near a river or lake or swamp. It was able to climb in the trees but favored bipedalism over climbing. I suggest protohominoid gave rise to three lineages. One lineage adopted arboreal life, perhaps because it was safer. These subjects acquired bodily specialization: long forelimbs and short hindlimbs (high humerofemoral index), hands for suspension, feet for prehension. They used their long forelimbs and short hindlimbs when moving on the ground with a new kind of locomotion where the second phalanges of the hands were applied to the ground: knuckle-walking. A second lineage coming from the bipedal protohominoid also climbed trees, perhaps for the same reason as the first lineage. These beings acquired climbing specializations but departed earlier and did not have enough time to complete their climber characteristics. This lineage gave rise to the Australopithecines whose lifestyle was probably little different from the ancestors of apes,

explaining their specific ape-like anatomy. Finally, a third lineage retained the ancestral bipedalism and acquired a very specialized morphology, especially for feet but also for limb proportions, with long hindlimbs and short forelimbs, etc. This lineage, which survived despite the lack of natural means of defense (e.g., horns, claws, etc.) because of the quality of its brain, led to more human-like forms.

The date given by Deloison for the existence of hominoid bipedalism is 15 MYA. Of course, this is speculation, but so is most of paleoanthropology. The time for the development of what we call the ape lineage is the Miocene (23-6 MYA); an extinction event occurred at the beginning or during this time period. Evidence for this is the Karakul crater in Tajik National Park in the Pamir Mountains, Tajikistan. Geologists and paleontologists recognize the diversity of animals that "show up" after large extinction events and for the Miocene one diversity was bipedalism recognized by at least 6 MYA, at the end of the Miocene, in the femur of *Orrorin tugenensis*. Orrorin's femur is fine-tuned for bipedalism and is more similar to that of modern humans, more similar than Lucy's (*Australopithecus afarensis*), especially in the neck's length that joins the head to the shaft and the positioning of the lessor trochanter. The fine-tuning of Orrorin's femur (morphological momentum) is evidence of a strong commitment to bipedality much further back in time, perhaps 15 MYA (or even further) as suggested by Deloison. In fact, *Homo ergaster's* femur is more similar to Lucy's than to modern humans (see Fleagle 2013: 380). This poses a problem because the long femoral neck of *Homo ergaster* is outside the range of modern humans, although the neck on the Orrorin femur is definitely within that range, and so is that of the chimpanzee and gorilla. *Homo ergaster* is likely, in some way, connected to the Australopithecine lineage.

Deriving efficient bipedal locomotion from a quadruped involves essentially a reworking of the total skeletal structure, and that an ancient ancestor to our protohominoid should possess some of these features, previously used for one purpose but are then repurposed for another function. The term for this is "exaptation" and is used to explain how different genomic entities may take on new roles regardless of their original function—even if they originally served no purpose at all (see Shubin 2020). One such candidate showing up in the Eocene (56-33.9 MYA) is *Tarsius eocaenus* dated to the Middle-Eocene c. 45-40 MYA. Rossie et al. (2006: 4381), referring to a facial fragment:

> This facial fragment, which is allocated to *Tarsius eocaenus*, is virtually identical to the corresponding anatomy in living tarsiers and differs substantially from that of early anthropoids such as Bahinia, Phenacopithecus, and Parapithecus. This new specimen indicates that tarsiers already possessed greatly enlarged orbits and a haplorrhine orinasal configuration by the time they are first documented in the fossil record during the middle Eocene.

Tarsier have some interesting characteristics. First, the foramen magnum is centered on the basal cranium to accommodate their method of locomotion in the trees. They are clingers and leapers, straightening their legs out when they jump from tree limb to tree limb, landing on their feet and then clinging to the limb. We have a critter, then, that could stand bipedally on the ground with its head facing forward. Niches get filled and can only support so many animals and it is possible that one of these Tarsier-like animals moved to the ground before becoming specialized for night hunting. The ground is not necessarily the safest place to be, especially at night, so our ground dwelling Tarsier probably never developed the large eyes. Its method of defense would be the ability to leap into the trees when necessary, but its food sources were ground based. It would only be a matter of time and cellular choice (epigenetic changes) to alter a physiology in line with Deloison's protohominoid. This is speculation, but at this point in time we need fresh ideas.

From the above we can see the probability of bipedal locomotion may extend back to the Eocene. In any case, with Orrorin we have a biped with a fine-tuned femur. We have no idea what Orrorin's foot might have looked like but we do have a curious footprint from Crete, dating to c. 5.7 MYA that shows "hominin-like" features which predates earlier footprints by over 2 million years (Gierliński et al. 2017). Six million years ago the Mediterranean Sea would have been much lower than today and Crete would have probably been accessible from the African continent. Much of the eastern Mediterranean was a migration route out of and into Africa during various periods of time between 7 and 2 MYA. Some anthropologists suggest that hominins might have evolved outside of Africa; anything is possible. Consider the complexity of all this, of who is who and what is what, we realize that we really don't have the necessary fossilized remains to develop clear, unambiguous conclusions as to human origins. The complexity factor suggests we are not dealing with random events leading to speciation and that there is an intense interplay between the environment and the cell (life), with the cell more of an active participant in its own evolution than previously considered.

Inbreeding and Human Evolution

Beginning around 2.5 MYA, right until the advent of sedentary agriculture 10 to 12 KYA, our ancient ancestors lived in small isolated groups that ranged over much of Africa, Europe, Asia, and the Americas – possibly by 130 KYA (Bower 2017). Population densities are difficult to predict. As small group animals, there are limitations on group size for basically two reasons: first, a specific territory, and food sources available, will only support so many individuals, and when that territory can't feed an increase in population, stress increases, and the group splits. Second, humans have information processing limitations first pointed out by Miller (1956) in the 1950s. When group size gets above 100 (see Pfeiffer 1982:191), and probably much before this, stress increases in the group because of information overload. Group stress creates a crisis and the quickest solution is withdrawal or changing one's environment; the group splits. There are dynamic sets of events occurring in any group, but the point is, population density can have the result of pushing groups apart, spreading the hunter/gatherers out into unknown territories— some adapt and survive, others do not. Over time, these different groups act as breeding populations where mates are exchanged. For the Neanderthals the genomics indicates that women migrated from group to group suggesting a patrilocal arrangement. We do not know if the erectus-types were patrilocal or matrilocal, but there must have been interbreeding.

Rupe and Sanford (2017; also see Sanford 2014 for a discussion of the weakening of the human gene pool) make a case for inbreeding resulting in certain morphological, phenotypic abnormalities, for example, when explaining the differences in cranial/facial configurations among the Dmanisi and *Homo floresiensis* finds. Although this may have been the case for earlier Hominids, recent research (see Besenbacher et al. 2019) suggests, by making comparisons with chimpanzees, gorillas and orangutans, human mutation rates have gone down over the past million years, so now there are fewer mutations than our ape relatives. "If the new mutation rates for apes are applied, the researchers estimate that the species formation (speciation) that separated humans from chimpanzees took place around 6.6 million years ago. If the mutation rate for humans is applied, speciation should have been around 10 million years ago... The reduction in the human mutation rate demonstrated in the study could also mean that we have to move our estimate for the split between Neanderthals and humans closer to the present" (*Science News*, January 22, 2019).

The questions would be, why have "mutation" rates changed, have "mutation" rates altered in the past, and what would cause such oscillations? If cells are thinking and making choices, should we still call these mutations or would the better fit be *endocellular* selection?

Going back to Dmanisi, the phenotypic differences we see among the erectus crania could be alterations called for by the cell in response to environmental stressors in different environments and/

or in response to sexual selection. The various configurations of crania found at Dmanisi could represent different families coming from different areas of Africa over a period of several generations rather than several millennia, explaining why they are found at a specific level. Dmanisi (located between the Black and Caspian Seas) likely represents a corridor through which many hominin groups passed as they moved east and west; there are likely many "bus station-like" sites to be found on that migration route. Returning to the crania, among all the Homo groups it is the cranium and face that seem to alter most rapidly over time rather than structures from the neck down (recall the dog cranium). So, if these are genetic abnormalities/mutations, why mainly the face? As long as the imbedded features (eyes, teeth, jaws, muscles) function properly, facial features could differ, even within the same hunter/gatherer group. Faces are like fingerprints; any configuration, as long as functionality is maintained, will work, and sexual selection may play a large part in the various configurations suggesting choice over chance.

Reconstructing Fossilized Remains

Let's say a person is murdered and left in a field in a temperate climate where there are surrounding trees, insects, scavengers, and varying weather conditions from hot to cold, from rain to shine. Unburied there will be virtually nothing left of the body within several months. Until very recent times when a hominid or hominin died he or she was not buried. Rarely circumstances occur where remains are well preserved after millions of years. Such circumstances would include falling into a swamp, or margin of a lake, and being rapidly covered with silt (*Homo ergaster*). The partial remains of the Taung infant (*Australopithecus africanus*) fell into a cave opening, because of the rather messy eating habits of an archaic leopard (*Dinofelis barlowi*), and were rapidly covered over. But this is generally not the case; fossilized hominid remains are rarer than diamonds. So, the researcher has to make the most out of what he or she finds; it makes this much easier if there is already a model in place for comparison.

We are equipped with wonderful imaginations; we are all biased—it can't be escaped, and we can fool ourselves into believing our story to be the correct interpretation. As mentioned, our minds are not equipped with a filter for right or wrong imagination.

An example of "imagination" is the specimen found by Tim White and company, *Ardipithecus ramidus*, dated using stratigraphy, index fossils, Potassium-Argon, and so on, to 4.4 MYA. The only parts of that specimen well preserved were the hands and feet, and these are not the hands and feet of a biped, perhaps on *Star Trek*, but not in the planet Earth hominid play book. To begin, Ardi's pelvis was in small bits and pieces, crushed beyond recognition. Owen Lovejoy used computer modeling to put the pelvis together, but this was programmed. There were 14 separate computer reconstructions and what programming specifications did Lovejoy eventually chose? Those of the *human* pelvis. The pelvis does not go with the feet and hands. This was obviously a climber, and the pelvis does not belong to a climber—the ilium is too short. Again, Ardi's pelvis was reconstructed by Lovejoy as being very human-like; experts in the field disagree (see Zorich 2010; also see John Hawks Weblog 2009). Because of this imaginative reconstruction, White then could assume that the foramen magnum (the orifice in the basil cranium) was centered rather than more to the back of the cranium, as would be the case of a chimp. The problem is, most of the base of the cranium is missing so it is difficult to know the position of the foreman magnum.

Further, the nuchal crest (also missing) was reconstructed to resemble that of Lucy or a chimp which doesn't necessarily point to bipedality. Moreover, the recovered right femur is missing the distal and proximal ends, but with an imagined bowl-shaped pelvis this creature must have had a locking out knee joint! A bowl-shaped pelvis and a locking out knee joint would suggest, to me at least, that Ardi was committed, and for a very long time committed, to bipedalism. Here is the problem. The pelvic reconstruction is not that of an animal who spent a great deal of time in the trees; the ilium is too short when compared, for example, to chimps and gorillas. It was imaged to be similar to that of a seasoned biped. So why the gorilla-type hands and feet? These in no way match a seasoned, committed biped. Also, when what remains of Ardi's cranium is compared to a chimp's there are many similarities. Ardipithecus is more than likely a quadrupedal ape, perhaps an ancestor of the modern chimpanzee.

I also have reservations about *Sahelanthropus tchadensis* (Chad Man) described in 2002 (see Dorey 2019) and dated to c. 7 MYA. Only the cranium was recovered and was compared to fossil animal remains radiometrically dated from other sites. The problem I have is the position of the foramen magnum – it is toward the back of the skull suggesting a more chimp-like locomotion. Although not everyone agrees about the importance of the position of the foramen magnum in bipedalism, it must play some part or all hominids would not have it centrally placed.

In 1944, teeth and mandible were discovered in Azmaka, Bulgaria and dubbed *Graecopithecus freybergi*. Dated to between 7.17 and 7.24 MYA it is considered by some (see Fuss et al. 2017) to be a hominin and the oldest member of our lineage found to date. If this is the case, then there may be some validity to assigning bipedality to Sahelanthropus.

And then there is *Homo ergaster*. This, according to the researchers, was a 9 to 10-year-old male who, apparently weaken from an abscessed tooth (septicemia or blood poisoning), fell into a swamp, and died. He was rapidly covered over around 1.7 MYA and recovered in 1984 by Kamoya Kimeu and Allan Walker. There are many similarities between ergaster and other erectus-types except one—the neck that connects the ball to the shaft of the femur. That neck would be a conservative trait, fine-tuned with the rest of the femur for bipedalism. However, being a conservative trait, that neck is long and appears outside the range of human variation. In fact, it is almost identical to Lucy's femur. On the other side, Orrorin of c. 6 MYA has a proximal femur almost identical to ours. One has to wonder if ergaster is a refinement of Lucy's kind and part of her lineage rather than the Homo lineage.

Dating Methods

The genetic/genomic information has opened a new vista in paleoanthropology keeping in mind that science is a process of discovery and sometimes we make initial assumptions with insufficient information only to find a fly in the ointment, so to speak. This leads us to radioactive dating methods, for example, Potassium-Argon, Uranium 238, and Carbon 14 dating, and why it's important to cross-check dates with stratigraphy, index fossils, tree rings, and so on, and keep an open mind. We have to address our initial assumptions about these methods. With dating methods, as with Neo-Darwinism, new dates and new data are dismissed in favor of the initial model (see Rupe and Sanford 2017: 269-306).

Potassium-Argon

The decay of Potasium-40 (^{40}K) to Argon-40 (^{40}Ar) occurs at a constant rate, or so we are told, with a half-life of 1.251×109 or 1.2524±0.0064 billion years. One of the key assumptions connected to K-Ar dating is that the material sampled came from volcanoes and the molten materials degas all argon present while in the molten state and before it solidified. In other words, the assumption is the samples would have no argon present once the material cooled and crystalized. This has turned out to be a false assumption (see Austin 1998; Snelling 1998). Because argon gets caught in the crystals, and not degassed, and is in the sample from the beginning, results will give an age much older than what it should be. For example, lava tested from Mount St. Helens, Washington, the dome of which crystalized in 1986, tested to between 340,000 and 2.8 million years old (see Austin 1998)!

Lucy was dated using the Potassium-Argon method yielding a date of 3.2 million years; Ardi was dated to 4.4 million years, and Orrorin was dated to between 5.8 and 6 million years ago. Either they are, or they are not the ages assigned. Major problems surface including the politics of who is right, along with fame and fortune, as it would appear there is "wiggle room" when interpreting results. This is especially the case where a model is presented, for example, Lucy (*Australopithecus afarensis*) as the base of the family tree, and Lucy's kind evolve into erectus-types, which evolve into Neanderthals, and eventually modern humans. This was the model 45 years ago, and it supports one narrative of human evolution; this is the Don Johanson-and-his followers' story. On the other side, initially at least, is Lewis and Richard Leakey's model that suggests the base of the family tree can't be Lucy because erectus-types were around at the same time as Lucy—the one could not have evolved out of the other.

The researchers, Johanson in Lucy's case, and White, in Ardi's case, after considering the various dates available (stratigraphy, index fossils, K-Ar, and so on), chose the date that came closest to their "desired" assumptions. So, how ancient are the remains of Lucy and Ardi?

Fission Track Dating

Fission track dating, used with certain minerals like zircon, involves the spontaneous decay of uranium-238 to thorium-234. In the process it emits an alpha particle, causing a trail of damage in the mineral. The age of the mineral can be determined, through the use of an electron microscope, by calculating how much U-238 was in the original sample and then counting the number of "lines" of damage in the material. Geologically, materials ejected from volcanos eventually solidify and initially, theoretically at least, there should be no tracks showing decay. So, for example, material is ejected from a volcano 3 MYA and some material lands on the ground. Within perhaps a day, week, or month later, one of our hominid ancestors suffering blood poisoning from an abscessed tooth, falls down on this very spot, and dies. Three million years later, a seasoned paleontologist, Dr. Erving Snodgrass, comes along, and just by luck, uncovers the remains of, "It looks like a hominid!" Soil is carefully collected from the layer, other faunal remains are collected, and so on. Zircon grains are extracted from the soils and subjected to fission track dating, and to his surprise, the dates vary widely from 1.5 to 150 MYA! Thinking rationally, he assumes that contamination has occurred and settles on dates closest to the model (4-3 MYA) favored by Dr. Snodgrass (see Gleadow 1980).

Carbon-14

We base life on carbon, an element formed in the compressed core of dying stars called red giants. The compression forces are strong enough to fuse helium nuclei (alpha particles) together to form carbon. There have been lots of red giants and they have produced massive amounts of carbon without which life as we know it would not exist.

There are three isotopes of carbon, ^{12}C, ^{13}C, and ^{14}C, with the mass number located upper left of the element. Nitrogen-14 (^{14}N) bombarded by radiation in the upper atmosphere gains a neutron resulting in ^{14}C, which is unstable. C-14 is then taken in by plants and animals in the form of carbon dioxide (CO_2) resulting in is what we call the carbon cycle. All life forms have acquired ^{14}C, at death the cycle is "broken," and the unstable isotope decays, thus changing the ratio of ^{14}C to ^{14}N. C-14 decays at a constant rate measured in terms of half-life, which is 5730 +/- 30 years. This means that every 5730 years half of the ^{14}C has decayed to ^{14}N, and in 5730 years half of that will have decayed to ^{14}N, and so, sort of like cutting a piece of thread in half again and again. In a good lab, dates as far out as c. 60 KY and as recently as c. 1200 CE (Current Era) can be obtained.

The US Government contacted Willard Libby, an American physical chemist, to determine the effects atom bomb testing was having on the atmosphere. Out of this he developed ^{14}C dating. His initial assumption was that the amount of ^{14}C in the atmosphere was constant and he calculated the ratio of ^{12}C to ^{14}C using Egyptian mummies and bread from Pompeii. Libby realized, however, that atmospheric levels of ^{14}C would vary over time according to geological activities, for example, volcanoes throwing out a lot of CO_2, solar radiation, industrial gasses, car exhaust, and certainly atom bomb testing. By understanding these variables, researchers have been able to refine dates. Also, dendrochronology (tree-ring dating) can be used to verify dates out to about 8000 BCE, for example, using the Bristlecone pine tree in California.

The other major issue is contamination. Assigning dates to materials sitting in museums prior to the smoking ban will usually show recent dates. Materials that were handled or touched with ungloved hands are also contaminated. Moreover, levels of ^{14}C in the atmosphere and oceans are different, thus dating materials that have been underwater for some time can be problematical.

Evolution and the Human Lineage

As discussed earlier, Deloison (2004) suggestion of a biped protohominoid existing c. 15 MYA in the Miocene is very intriguing. As mentioned, we need fresh ideas regarding our evolutionary history as we have been handicapped by forcing new data into old models, the main one being that the genus Homo evolved from the Australopithecines. It would appear, however, that Homo and the Australopithecines were walking around the same landscape during the same time frame—the one could not have evolved out of the other. If we have a protohominoid/biped wandering around 15 MYA, perhaps we've been looking in the wrong place. We have Miocene deposits in Kenya (e.g., Rusinga Island) where Allan Walker found Proconsul, and then there are areas in eastern Kenya, Saudi Arabia, Pakistan, and China. Excavations for the most part have been carried out mainly in Africa, again prompted by the belief that we descended from Australopithecines, which we assume were confined to Africa. According to Rupe and Sanford (2017: 342):

> Figure 3. The paleo-community openly acknowledges that hominin record (the actual data) does not reflect an ape-to-man progression. Instead, there appears to be a clear separation between the ape and human type. There is a lot of diversity

within the ape types and a lot of diversity in the Homo types (with many variants now extinct). There is also branching within each group. Yet we are not seeing a fossil trail connecting these two very distinct groups (ape and man) via a series of intermediate forms. A more accurate representation of the data is with multiple branching trees (reflecting adaptive radiation over time), rather than a single tree with a single common ancestor of ape and man.

Along with this perspective, Rupe and Sanford (2017: 342) offer a diagram representing their proposed evolutionary relationships. At the bottom of the graph they position the gorilla-like ancestor evolving into the modern gorilla, with connections to *Sahelanthropus tchadensis* and *Australopithecus afarensis*, with the explanation that some afarensis bones are like the modern gorilla. Next in line is a chimp-like ancestor evolving into the bonobos and chimps, with connections to *Ardipithecus ramidus* and *Australopithecus afarensis*, with an explanation that some afarensis bones are chimp-like. Then there is the Orangutan ancestor (possibly *Sivapithecus indicus*) evolving into the modern Bornean and Sumatran varieties.

Next there is the Australopith-like ancestor, now extinct, with connections to *Paranthropus aethiopicus*, *robustus*, and *boisei*. Finally, there is the *Homo sapiens* ancestor, a biped, that evolves into modern humans, with connections to erectus-types, Neanderthal, naledi, Denisovans, and the Hobbit (*Homo floresiensis*). This model differs from models proposed by Johanson, White, and others, and more closely fits with the narratives presented by Deloison and the Leakeys. Obviously, both can't be correct and we need more specimens from earlier Miocene deposits.

Topobiology and the Hominin Knee

Topobiology, evolving from the mathematics of topology, is particularly interested in how contiguous cells assess their local environment and adjust their patterns of gene expression to match their position in the developing embryo. Topobiology is thus concerned with cell surface proteins that allow cells to sense their surroundings and transmit control signals to the nucleus. In some cases, cells respond to these signals by radically changing their "lifestyle" and migrating to new locations in the embryo. My interest in topobiology centers of how cell surface proteins, including lectin-carbohydrate interactions, sense their surroundings and then signal the nucleus. Feedback from the nucleus can lead to alterations in the systems to which the cells belong. For example, contiguous cells/tissues communicate in terms of sculpting joint surfaces, the head of the humerus and the glenoid fossa, or the interaction of the proximal tibia and distal femur, during growth phases and through repetitive use at maturity. Out of this, phenotypic variations emerge, usually to fine tune a system. Can these modifications pass on to future generations? As there are levels of communication between cells including germ cells, some of Lamarck's more controversial examples might have some validity.

I approached the subject of topobiology in *Clinical Anthropology* (1996), mainly regarding brain functioning, intelligence, and information process using Edelman's (1988) perspective. Edelman's work, in my opinion, has not received the attention it deserves. Combined with more recent research, I see another important part of the biological evolutionary process—especially with data showing cells can think and make decisions.

Another term for topobiology is "morphogenesis" and, along with cell differentiation and cell growth, is part of what is today called evolutionary developmental biology (evo-devo). (For the technical data connected to my discussion see Gilbert 2000; Kouros-Mehr and Webb 2006; Montevil et al. 2016; Bidhendi et al. 2019). I have expanded my research to include the ankle, knee joint, and pelvic girdle. Taking Lamarck's basic model seriously (acquired characteristic, or the organism making alterations in the face of environmental stressors), I began to explore what it would take to convert a chimp's knee into a hominin knee and vice versa. What I discovered is this:

there are lots of alterations that have to occur but, because they are part of a *modular* system, changing just a few parts can lead to bipedalism.

Crawling

Humans are the only primate that crawls as a second step in building its upright, bipedal form of locomotion. The first step is lying on his or her back and leg kicking. Apparently crawling is not unique to humans. According to recent finds in Argentina (Nield 2019), of a dinosaur *Mussaurus patagonicus* (Mouse-lizard) that lived c. 200 MYA, show evidence that this ancient creature, from the late Triassic and early Jurassic, crawled like a human baby before it walked.

For the human infant, with short arms and long legs, there isn't much of an alternative if the infant wants any mobility at all (although infants pull or push themselves around while lying on their back or stomach). The characteristics of a chimp's mobility, i.e., quadrupedal knuckle walking, are established in the womb, so at birth the Chimp hits the ground running. All human infants are born prematurely, a condition brought about by a bowl-shaped pelvis that restricts pregnancy to around nine months because of brain growth. In short, all human infants are born premature, which has the unintended effect of altering group dynamics and relationships between males and females, if the genome is to survive.

Let's begin with Lucy (*Australopithecus afarensis*). The humero-femoral Index (length of humerus X 100 divided by length of the femur X 100) shows Lucy's arms and legs are closer in length to human than that of a chimp. This suggests that Lucy's kind (the Australopithecines, and I will include the Paranthropus varieties) might have crawled before they walked. Human infants crawl out of necessity; with short arms and long legs, the distance between knee and acetabulum corresponds closely to the length of the arms. Keep in mind that for human infants the acetabulum has not fused. The ilium, ischium, and pubis all fuse at 7-9 years, but the acetabulum, or replacement of the triradiate cartilage, fuses around 20-25 years. For the human infant, upright walking is not a possibility (usually) until after 12 months of age. Let's look at some structures involved.

When comparing the chimp with the human knee, one can quickly see that it is a matter of lengthening or shortening specific muscles and bones; nothing new needs to be added. However, there are several muscles specific to the chimp that humans do not possess, i.e., Psoas minor, Scansorious (stabilization of the pelvis), Abductor minimus, Sichiofemoralis, Opponens hallucis, Opponis digiti minimi, Ontrahentes pedis (flexion and abduction of phalanges of the foot). The chimp pelvis needs stabilization because of the chimp's side-to-side gate during quadrupedal locomotion. Also, the chimp has a grasping foot, thus the need for muscles to control gripping. One has to wonder if our human ancestors ever had these muscles and that depends on the common ancestor with the chimp. As mentioned earlier, there is evidence of a human foot found in Crete dating to c. 5.4 MYA, and if this is what it appears to be, the human foot was around long before 5.4 MYA. Lucy (*Australopithecus afarensis*) has a foot with an arch, a fine-tuning for bipedalism, but a slightly opposed big toe and an inner ear, like a chimp, that is not designed for bipedalism; Lucy dates to 3.2 MYA. But we know her kind was around long before 3.2 MYA. Probably she was contemporary with *Ardipithecus ramidus*, meaning the one did not evolve out of the other. So, if we have a human foot at 5.4 MYA, it suggests that the human lineage does not include Ardi or Lucy.

Other differences between a chimp and human are the length of the femur, the splaying of the ilium, and the configuration of the knee joint. Let's start with the knee joint. The articular surface on the tibia for the medial condyle of the femur of the chimp is round, while that on Lucy and *Home sapiens* exhibits an anteroposterior elongation—more so on the human. The proximal end of the human tibia is much "beefier," with the total platform taking on a squarish configuration. This is for weight bearing showing the need for such a large platform, especially when running.

A genomic comparison between chimp and human hip and lower leg bones and muscles would reveal that the differences are not because of random mutations but the addition or subtraction of methyl units from DNA coding thus affecting the expression of these body parts. Considering this, I predict that the human genome will not contain derelict coding sequences for the muscles listed above. If this is the case, it would show that the human lineage, as suggested by Rupe and Sanford (2019), might have been bipedal right from the beginning, which is anywhere from 23 to 15 MYA. This brings us back to the crawling behavior of human infants.

Crawling behavior of human infants begins around 7 months and usually ends around 10-12 months, at which time they begin to walk. Balance and coordination are usually well installed by 14-16 months (there are certainly differences in timing, but that need not delay us here). There are at least two good reasons infants don't skip crawling and go right to walking. The first is "they don't have a leg to stand on." The epiphyses (distal and proximal ends) of the femur and tibia and muscles and ligaments are too immature for weight bearing until 10–12 months after parturition.

The second issue is neurological and balance. The inner ear (semi-circular canals, part of the vestibular system) has to receive competent signals/feedback from the phalanges of the foot, the ankle, the knee joint, hip joint, and visual cues or it cannot send correct signals to the muscles that control balance. If the nerves and muscles are not mature enough, walking can be difficult at best. Crawling serves many purposes. It helps to create upper body balance and strength, and it aids in the knee's construction and hip joint and perhaps the upper arm (head of humerus) and scapular joint (glenoid fossa) as well. All joints engage in communication across joint surfaces with osteoclast cells sculpting and remodeling bone and associated connective tissue. These cells are controlled by the hormone estrogen. Communication is not just at the joint surfaces but extends throughout the body as one hand does need to know what the other hand is doing, so to speak.

Crawling seems to be an important part of an infant's physical/neurological development or it wouldn't be universal. Here is a possibility. Crawling (contact with the outside world), in conjunction with the weight and movement of the body over the pelvis and thighs is "telling" the bones to configure in particular ways, for example, the size and shape of the proximal head of the tibia to accommodate eventual weight bearing on the distal end of femur. Again, weight on the proximal end of tibia possibly has some bearing on the anteroposterior elongation of the tibial platform and the anteroposterior elongation of the medial condyle fossa on the platform. As the human femur is angled in, the extension of that fossa would keep the medial condyle of the femur more securely in place especially when running. We see a similar situation on Lucy's proximal tibia and, although not as elongated, surely indicates she was a biped. On chimps that fossa is circular. Did Lucy begin her ability to walk on two feet by first crawling? Could this be a generalization for all bipeds? Is there any difference in development with modern children who do not crawl?

Conclusion

The study human ancestry has stalled for several reasons, mainly because the availability of specimens; fossilization only occurs under the most unusual circumstances and then finding those fossils many times is just pure luck. Moreover, primates, as social animals, go for quality over quantity and compared to the reproductive physiology of dinosaurs, for example, we have few offspring. We have opted for "culture," care of our young rather than relying on luck or raw instinct. Some reptiles are social animals; many birds fall into the same category. But others, like sea turtles, socialize little with their young. Mother turtle walks up on a specific beach when it is time, digs a hole, deposits her eggs, and heads back to sea. Once out of the shell (baby turtles are about the size of a quarter) the baby turtles head for the sea—they are on their own. Most don't make it, but there are many eggs and lots of little turtles. Another issue that has, perhaps, led us astray, is the image of an hunched over, ape-like animal, evolving into a biped that would eventually "turn into" an Australopithecine, from there to an erectus-type, and so on. Our ancestry is muddled by models.

A third issue is the belief that all these evolutionary changes have occurred randomly and that organisms are at the mercy of random mutations and natural selection. From at least the time of habilis (c.2.5 MYA) our lineage has attempted to alter the effects of the environment through technology and the modifications of our surroundings (macro-changes); I see this as analogous to the cell making changes at the micro-level. One has to keep in mind what is trying to survive; it isn't the prototype, the human, for example, it is the instructions for building life. And just like our survival depends on our taking an active role in response to nature, the cell does the same—it is an active participant in its own survival. Any evolutionary model has to take into consideration the cell's ability to think and make decisions, not only at the time of environmental changes if not before. The human knee joint, like any joint surface, is suggestive of this thought process. From an evolutionary perspective an ape like the gibbon just seems to show up in the mid-Miocene with long arms for brachiation; knuckle walking quadrupedal apes, like the chimp and gorilla, just seem to show up in the Miocene. There aren't too many discussions in the anthropological literature questioning the origins of brachiation, or, for that matter, knuckle walking. Perhaps our branch of the tree, generic bipeds, might have "just shown up" in the mid-Miocene.

CHAPTER 8

Conclusion

WE need models as a starting point for understanding what we are and our origins. Many times, however, we assume our models are correct and thus data that corresponds with our initial beliefs is accepted, and that which is contrary is often ignored or rejected. The Neo-Darwinian model is a case in point. Moving past the model presented by Darwin, the Neo-Darwinists maintain the assumption that life evolved incrementally and randomly over long periods of time. Once DNA was better understood in the 1950s, knowledge of amino acids and base pairs, this seemed to bolster the idea that random mutations plus natural selection plus time equals new species. In this outdated model, everything has to be random; there can be no other explanation for the origins of the universe, and the life forms within it. This being the case, the universe and life have no meaning or purpose. Statistics and Behe's irreducible complexity, however, threw a wrench into the Neo-Darwinian gears. Instead of questioning their own assumptions, the Neo-Darwinists did their best to attack and demonize Behe and others presenting views contrary to theirs. This also led to models, such as the multiverse, in a weak attempt to "prove" that, if there are infinite universes, then at least one—the one we live in, could have randomly "created" life as we know it. Positing multiple universes, however, gets us no closer to understanding the origins of this universe, let alone the origins of organic life forms. Just as the Neo-Darwinists cannot prove that random mutations lead to new species, we cannot prove the existence of a multiverse either. Yes, we have mathematics that lead us in that direction, but the math always starts with assumptions, and we can't be sure the assumptions are correct. In my experience, you can probably formulate math, along with computer programs, to prove anything you want.

As stated in the previous chapters, there are many mechanisms, not acknowledged by the Neo-Darwinists, that lead to new species and these include epigenetics, symbiosis, and hybridization. Further research into quantum biology and topobiology hold promise in revealing other avenues of biological evolution. Neo-Darwinism assumes that random mutations and natural selection result in new species. This is an idea, a hypothesis, and does not represent the process. An example of a process expressed mathematically is that of gravity. Newton came up with the formula **$F = G([m_1 m_2]/r^2)$**, which specifies **F** as the force of attraction between two bodies, **G** as the universal gravitational constant, **m_1** as the mass of the first object, **m_2** as the mass of the second object, and **r** as the distance between the centers of each object. The Neo-Darwinians have nothing like this to measure or test their position; they assume that the random mutation is the process—no math required. The concept of random mutations and natural selection as the model has been cited repeatedly in book after book, and because it's touted by those in "authority," such as Richard Dawkins, it must be true. As I mentioned, if a person spends a great deal of time, energy, and money promoting an idea, new perspectives are likely rejected. This mental state is called cognitive dissonance.

But there is another issue that has kept the Neo-Darwinian model alive and that is social Darwinism, or the belief that those in power or with money are more highly advanced or evolved than others. Therefore, they should rule over these inferior people or even eliminate them. This led to eugenics, socialism, the social democrats (Nazis), and communism. We can see this Marxist attitude alive and well in most areas of the world. The Democrat party (not democratic party—they have little to do with democracy) in the US is bound to this idea. For the Democrats, they are the all-knowing elite and the rest are scum. Marxism also supports atheism and the idea that life has no meaning—there is no purpose to life. I don't know if there is a god or not, nor do I know if there is a purpose in life except to slow entropy. Perhaps that is the purpose. That is, the purpose of life is to slow entropy long enough to develop an understanding of the universe and what life is all about, for without life the universe is meaningless, as only consciousness and intelligence can confer meaning.

Marxism, Elitism, and Atheism

Staying with atheism, there are authors actively prompting atheism with some very interesting arguments. Why am I getting into this? Belief in an intelligence lurking behind the universe was the "peril" that Darwin and his followers were fighting against. To "disprove" the existence of intelligence that designs, orchestrates, and builds the universe, they point to poor design features thus limited intelligence (see Hafer 2015). Hafer (2015: 174) makes the following statement:

> On the other hand, the basic mechanism of evolution is mutation, followed by natural selection... That's how evolution works! That's how you get a new species, or even several new species, from one original species. Mutation is one of the fundamental drivers of evolution.

Hafer goes on to say (2015: 174), basically, that mutations, although the "fundamental drivers of evolution," should not exist if the "designer" was intelligent. The first statement echoes the Neo-Darwinian dogma that mutations plus natural selection plus time equal new species. This is the same as saying that the apple fell from the tree because of gravity without explaining the details. The apple does not fall to the ground because of the word "gravity." The apple fell to the ground because of the inherent rules, available at the beginning of the universe, of attraction of particles with mass. We have the math for this, and the math allows us to go to the moon.

"Mutations" are the source of new species, according to Hafer, but I'm not sure what she means by mutation. I assume it is a random alteration of base pairs. What I would like to see is the process, in detail, of how random chance can build a functioning protein, an enzyme, or a cell. You just can't invoke the belief in "mutations" as the mechanism, as this does not differ from invoking God. Show me how the mechanism works—let's get into the math, physics, and chemistry. I've seen none of this; I've seen mutated fruit flies but the end point has never been a new species, only less fit monsters, with legs growing out of their heads and so on. Hafer doesn't seem to be aware of the possibility of a generic set of rules for constructing life, with nothing outside of the cell, bacteria, and viruses predetermined. These rules are life's default system. What Hafer is referring to are unintended designs through repurposing existing structures (see Shubin 2020).

Hafer (2015: 6) further claims that "ID folks are a political lobby," and then compares ID (Intelligent Design) to the tobacco companies, a political maneuver to demonize ID. I hate to break this to you, but Neo-Darwinism is likewise a political movement connected to eugenics, Marxism, Nazi Germany, communism, and the torture and murder of millions of innocent people. Most of the intelligent design-oriented scientists that I know believe in evolution—animal and plant forms change. The disagreement lies in the *process* or *processes* of change, and for the ID scientists, random doesn't get it done. As I mentioned, subscribing to ID does not translate into a belief in the existence of a god. Many of the scientists are cautious with this only saying, "there is some kind of intelligence behind this." And yes, there are scientists who are true believers.

As an anthropologist, grappling with this for many years, I've concluded that we don't know as much as we think we do, all models of origins of the universe and life fall short of testable explanations, and therefore it would not be good science to sideline any position. However, it doesn't make a bit of difference to me whether there is or isn't intelligence behind all this; I want to understand the design, the actual process, of how you get from inorganic materials to organic materials and life.

Hafer (2015) confuses ID with religion; they are not the same thing, although many religious groups hijacked ID to prove the "truth" of their narratives, which shows they don't understand ID either. Intelligent Design likewise does not suggest that everything, every plant and animal, is preset and coded right from the beginning. Rules were set right from the beginning, but the original rules of the universe are generic, not specific to every life form. Once you know how to build a cell, enzymes, DNA, RNA, etc., these can arrange in a multitude of ways. All life forms are composite modules (the analogy used are Lego bricks) and with them you can build many life forms—they are variations on a theme. Once again, Hafer (2015), as a Neo-Darwinist, falls for the trap of believing that random mutations, plus natural selection, plus time, equals new species; this is not the process, but instead, and idea or hypothesis that has never been tested or proven correct.

Congruent Construction

Neo-Darwinists have demonized ID to the point where any mention brings forth the God image, an image that apparently strikes fear in the hearts of devoted Neo-Darwinists and liberal politicians. To avoid this and not attach myself to theistic belief systems I propose (see Chapter 1) the term "congruent construction." By this I mean, and only mean, that inorganic, organic, and organic life systems are just too perfectly constructed, with all the parts in place, to have come together in any manner that could conceivably be considered random. My concern is, how do you turn organic material into organic life forms? I don't have the answer to this question and, in fact, we may not be asking the right question. My position is, all the answers are in the quantum realm—if we can ever understand the order of that domain.

Genetic Dormancy

I mentioned genetic dormancy in Chapter 4, that is to say, during catastrophic events, for example, Snowball Earth just prior to 1.7

BYA and then again c. 700 KYA, that coding sequences might go dormant but still be active within the quantum wave. In this state I suggest that thinking and decision making could occur to prepare for the environmental conditions prevailing after the event, thinking ahead as evidences in the Dutch Hunger Study. This might explain the development of eukaryotes emerging after the first Snowball Earth and then the explosion of life forms during the Cambrian c. 550 MYA. This is only speculation. However, the wave form is not static; there is continual "movement."

Measurements have revealed numerous particles emerging from the wave form, which come under two general categories, three generations of fermions composed of leptons and quarks, and interacting forces or carriers called bosons. The first generation of quarks are up and down, we term the second generation, charm and strange, while we label the third generation top and bottom. For leptons, the first generation is composed of electron and election neutrino, the second, muon and muon neutrino, and the third generation, tau and tau neutrino. For the bosons we have gluon, photon, Z boson, W boson, and Higgs or H boson. The point: obviously there is a great deal of order existing within the wave form or, when measured, these arbitrarily named particles would not exist. Whether there are other particles remains to be seen. The question is, what is the wave form, that manifest as particles, doing during a dormant phase? With the complexity we see around us, I have difficulty imagining a static wave form. I think it reasonable to assume that eukaryote coding sequences in the wave form can rearrange sequences and create new possibilities. This also suggests that complex (irreducibly complex) genotypic or phenotypic changes only occur in the wave form. Thus, for example, the differences in a quadrupedal vs bipedal knee joint do not occur after the wave form has collapsed. Individual alterations to coding sequences may occur during the collapsed phase (e.g., mutation of a nucleotide from C to T, or A to G) but large changes, because of their complexity, occur in the wave form during dormant phases. This may explain the existence of new life forms after a catastrophic event. Again, this is only speculation.

Hafer and Irreducible Complexity

Hafer's (2015) position is basically about overturning intelligence in the development of life forms on this planet. Within her discussion she attempts to invalidate irreducible complexity by comparing blood clotting mechanisms found in various animal species, including sea cucumbers, sand dollars, starfish, and so on. The impression I received from this is that various clotting systems, or systems for preventing fluids from inadvertently leaking out of the various species, somehow make up the parts of the mammal blood clotting system. After reading this I imagined a horseshoe crab humping a sand dollar and begetting hundred-dollar bills! I guess I don't understand her brand of biology. Or maybe zoology and genetics are taught differently in Britain. (After all Dawkins' position is highly praised among the British, including lots of awards for academic excellence.) She even suggests that the blood clotting of a horseshoe crab is more effective than that of mammals, showing the "Designer" made a dumb move in not giving such a system to us. She encapsulates her analysis by saying (2015: 95):

> So, it has been demonstrated that blood clotting systems are entirely reducible, since reduced blood clotting mechanisms are found throughout the animal kingdom in many different living animals.

To begin, this is not a logical statement. I'm not sure what "it" refers to, that is, who or what demonstrated this. Within her example, she does not demonstrate blood clotting systems are "entirely" reducible. Her argument relies on the fact that different animals have different systems for preventing precious fluids from leaking from the body. We should expect this; some are simple and some are complex. But all are irreducibly complex; remove any part or programmed behavior and the individual systems don't work. Just because an animal has a "blood" clotting system simpler than ours doesn't make it reduced. Because we find a different system within different animals does not mean they somehow all miraculously came together to produce our system. There is no way statistically that blood clotting systems could have developed randomly (or incrementally gained from a group of diverse species—the statistical probability of this is beyond reason!) regardless of what animal we are considering. Hafer's position is one of desperation, an attempt to save the undermining of Neo-Darwinism and atheism. I don't think the Neo-Darwinists understand what random means. It means *haphazard, non-specific, arbitrary, unsystematic*. Random, by definition, cannot build systems. Remember, you can alter any part of the genetic code as long as you follow the rules. Random mutations do not follow the rules and that is why they kill, lead to ill health, or are mostly neutral. The one thing we cannot prove about irreducibly complex systems is how they developed; my best guess is they assembled in the wave form. Again, show me, with our materialist science, math, chemistry and physics, how random mutations can create new species and I'll be a believer.

Morality and Religion or Outgrowing Dawkins

Contrary to popular belief, religion and morality are not the same thing. Morality is an important issue for the Neo-Darwinists for they believe that morality is a social construct because concepts of morality differ from culture to culture. This challenge to morality, although around for a long, long time, burst forth in the 1960s in the "liberation" movements (which today we called liberalism—John Locke did not "invent" modern liberalism) where individuals wanted to be "free" to "do their own thing." There seems to be four major "things" individuals wanted liberation from: rules of sex, rules regarding drugs, freedom from work, and dress codes. There are more than these, but these were and are the most conspicuous. In a phrase, individuals wanted to be free of rules created by others. And who did many of these individuals grow up to be? Yes, professors, lawyers/politicians, and pseudo-journalists. When I was growing up, and that was prior to the 1960s, liberal meant open-minded, or perhaps you used a "liberal" amount of salt. Liberals today are anything but open-minded; I'm not sure what they think of salt. Morality does not stem from religion; it stems from social necessity. Certain behaviors upset social process and these include lying, cheating, baring false witness, physically abusing (including rape), and murder.

Morality has two sides to it: behaviors that tangibly affect others, and intangible behaviors that do not. Tangible behaviors center on lying, cheating, stealing, physically injuring or abusing others, and murder. These are tangible morals because they directly affect others, disrupt social functioning, and can lead to systems failure. In other words, these behaviors destroy trust; it is difficult to work with people you distrust.

Lying and distortion of facts is the mainstay of politicians, the techniques of which are presented and honored at prestigious law schools; some of the most prominent politicians are lawyers. The lying and distortion has gone on for so long we consider it normal for politicians—we even make jokes about such behavior. "How do you know when a politician is lying? His or her lips are moving!" Such jokes seem to make the behavior unlikable but acceptable. These tangible moral values have become confused with intangible behaviors, for example, displaying one's self in public or consenting adults desiring same sex encounters. You might be interested in looking, or feel disgusted with homosexuality, but the behavior has no tangible effect on you. Remember, your emotions are your own; no one can give them to you (see Rush 1999).

All cultures have sexual restrictions, probably originally designed to reduce stress in system relationships. Sexual behavior is so emotionally charged (after all we are animals) it has to be controlled. No, we are not like bonobo chimps! Who we have sex with is also an issue. Somehow our ancestors knew, and I don't think it coincidence, that inbreeding can negatively affect the physical health of the group. So, stress and inbreeding are two possibilities why there are sexual restrictions. But in the larger picture, Mother Nature doesn't care who or what you have sex with as long as there are enough males and females getting together to bring in the next generation. Homosexuality is not the norm—it is only a possibility. Back to morality and my point.

Because most of the Neo-Darwinists I've read to date are atheists, believing that morality is a social construct, and because morality is defined in different ways in different cultures, one can justify lying, cheating and stealing, abusing others, and so on—if the cause justifies it. A perfect example of this has been handed to us by the Democrat party in the US with the construction of the Steele Dossier, concocted by Democrat operatives and politicians, including foreign (British and Russian) operatives, to get rid of an elected president; this was a coup, now called Obamagate, orchestrated by the Obama administration. The Democrats believed they could get away with this because Obama, over the course of his 8-year administration, was able to criminalize the IRS, the FBI, and the Justice Department in an attempt to turn the US into a totalitarian state no different than Russia, China, Iran, and so on. You shouldn't be surprised by this statement; Democrats are social Darwinists, Marxist in orientation. Democrat politicians are a danger to the rights of the individual, they are corrupt, and an insensitive group of people. After all, it was the Democrats who created and maintained racism in this country by establishing and funding the Klu Klux Clan until the mid-1980s, along with the Jim Crow laws. I realize this is political and not the place for such dialogue, but it is a good example of where lying, cheating, stealing, killing, and emotionally abusing disrupts social functioning and has little or nothing to do with religion.

At the forefront of the atheist dialogue is Richard Dawkins with his latest forage into the human philosophical condition, *Outgrowing God: A Beginner's Guide* (2019). His main thrust is directed at organized religion, Judaism, Christianity, and Islam. I don't have any problem with his criticism of these traditions (see Rush 2008, 2011).

As noted earlier, Yahweh, God the Father (Yahweh brought over to the Christian condition—not Jesus), and Alilah or Hliliah (Allah) are demons, demons that, by analogy, represent the tyrants connected to the origins of these traditions.

Dawkins attempts to define atheist, pantheist, and agnostic, with atheism the only proper way to think. In Dawkins' world, it is even inappropriate to suggest you don't know if there is a deity behind the universe or not – he considers this a "feeble" excuse substituting for truth (Dawkins 2019: 12). He misinterprets the meaning of messiah (2019: 20) and the connection to anointing oil, the Old English meaning of "gospel" or God's spell (2019: 21), and he is unaware that Nazareth (2019:31) was not a geographical place until mythically constructed as such by confused priest poets. Nazareth was originally a reference to the Nazarenes, a Jewish sect whose membership included our mythic heroes Samson and Jesus. The exodus was not out of Egypt (2019: 50-55); the exodus was out of Babylon (see Rush 2008, 2011). More recently we have discovered that Jesus was an experience (and not a living, breathing person) obtained by consuming the *Aminata muscaria* or Holy Mushroom and most likely descending into the wave form "where all secrets will be revealed" (Rush 2011). Most of these critiques have been in the literature for over 150 years and they haven't done much to persuade individuals and groups to abandon God! What Dawkins really misses is Judaism, Christianity, and Islam are fascist political systems specifically designed to control people using the fear of God and God's punishment to keep the masses in line. This isn't much different from university departments discouraging any view that differs from the Neo-Darwinist's model by shunning, demoting, or destroying the reputations of those who don't share their beliefs.

Dawkins (2019) spends some time talking about morality and how the expression of morality changes from generation to generation. One problem I see is that human groups, out of necessity, have to discourage certain types of behaviors because they disrupt social functioning. Thus, as mentioned earlier, there are certain moral dictates that have evolved out of social necessity. Lying, cheating, theft, physical abuse, bearing false witness, and murder are disruptive forces in all cultures. I get the impression from Dawkins that, because morality is a "cultural construct," you should be able to act and do what you want. If I am misreading his position, he needs to explain to me what he considers moral behavior and its origins, as it doesn't come out of what we call religion.

Returning to definitions, atheism is, like Neo-Darwinism, faith based. Dawkins has faith, but no proof (using materialistic science) that random mutations produced the life forms on this planet. Dawkins' gods are random mutation and natural selection and their magical ability to transform inorganic matter into organic matter and organic matter into life forms and speciation. He and the other "bitter clingers" invoke "random mutation" without offering proof. Insisting that random mutations and natural selection are the sole moving forces in life's development and maintenance does not differ from believing in life's creation by Yahweh, Allah, or any of the gods mentioned in his text. Creation myths contain common elements, magical events, brought about by magical entities, who have special people (humans), founding fathers, represented by heroes who mirror and fight to preserve the myth. Heroes always have "enemies" or obstacles to cross, and magical help is always available. So, what we have in Neo-Darwinism is Random Mutation as the creative force in the universe—we can call this force Ran-Mut. Ran-Mut communicates with special people, for example, Darwin, Huxley, Marx, and so on, with Dawkins and others as loyal disciples repeating the mantra, "Ran-Mut, Ran-Mut, Ran-Mut," as the creative force in the universe and the royal road to new species—with the help of natural selection, of course. I am also left with the impression that, if you don't believe in random mutations, and that life has no purpose, you must believe in ID (Dawkins 2019: 174). You only have one choice, you must accept random mutations as the driving force in life with no reference to intelligence beyond the human animal. Once you realize that we are an accident and there is no purpose to anything, you grow up, mature, and put childish things away. I personally think it is time for Dawkins and his followers to outgrow the Neo-Darwinist position and put simplistic notions away. As Margulis comments (2003: 201-202):

> The Neo-Darwinists' inventive literature and valiant attempts to unite the genetic stability in the unblended mixture of Gregor Mendel's factors to the gradual evolutionary change promoted by Darwin's natural selection were as brilliant as they were incorrect. The hegemony of R. A. Fisher, J. B. S. Haldane, and Sewall Wright is gone forever, and their latter-day saints—Richard Dawkins and J. Maynard Smith, or at least their students—will have to learn something about chemistry, microbiology, molecular biology, paleontology, and the air.

Bibliography

Arnold, M., Brother, A., Hamlin, J., Taylor, S., and Martin, N. 2015. "Divergence-with-Gene-Flow – What Humans and Other Mammals Got up to." In, N. Gontier (ed.), *Reticulate Evolution: Symbiogenesis, Lateral Gene Transfer, Hybridization and Infectious Heredity.* NY: Springer, pages 55-295.

Austin, S. 1998. "Excess Argon within Mineral Concentrates from the New Dacite Lava Dome at Mount St. Helens Volcano." *Creation Ex Nihilo Technical Journal* Vol. 10 (Part 3).

Ayarpadikannan, S. and Kim, H. (2014). "The impact of transposable elements in genome evolution and genetic instability and their implications in various diseases." *Genomics and Informatics* 12: 98-104.

Axe, D. 2000. "Extreme Functional Sensitivity to Conservative Amino Acid Changes on enzyme Extractors." *Journal of Molecular Biology* 301:585-595.

Axe, D. 2004. "Estimating the Prevalence of Protein Sequences Adopting Functional Enzyme Folds." *Journal of Molecular Biology* 341:1295-1315.

Axe, D. 2017. *Undeniable: How Biology Confirms Our Intuition That Life Is Designed.* NY: HarperOne.

Bae, C., Douka, K., and Petraglia, M. 2017. "On the Origins of Modern Humans: Asian Perspective." *Science* 358, Issue 6368.

Barbieri, M. 2018. "What is Code Biology?" Biosystems Vol. 164, February, pages 1-10. Barry, G. and Mattick, J. (2012). "The role of regulatory RNA in cognitive evolution." *Trends in Cognitive Science* 16: 497-503.

Bateson, G. 1972. *Steps to an Ecology of Mind: Collected essays in anthropology, Psychiatry, Evolution, and Epistemology.* Chicago, IL: University of Chicago Press.

Behe, M. 2019. *Darwin Devolves: The New Science About DNA that Challenges Evolution.* NY: HarperOne.

Behe, M. 2007. *The Edge of Evolution.* NY: Free Press.

Behe, M. 1996. *Darwin's Black Box: The Biochemical Challenge to Evolution.* NY: Free Press.

Berlinski, D. 2009. *The Devil's Delusion: Atheism and Its Scientific Pretensions.* NY: Basic Books.

Besenbacher, S., Hvilsom, C., Marques-Bonet, T., Mailund, T., and Schierup, M. 2019. "Direct estimation of mutations in great apes reconciles phylogenetic dating." *Nature Ecology & Evolution*, 3, pages 286-292.

Bierman, D. 2008. "Consciousness Induced Restoration of Time-Symmetry (CIRTS), a Psychophysical Theoretical Perspective." *Proceedings of the PA Convention*, Winchester, England, August, pages 13-17.

Bodovitz, S. 2017). "Consciousness of Continuity in Time," In, *How Consciousness Became the Universe: Quantum Physics, Cosmology, Neuroscience, Parallel Universes,"* pages 309-315. *Defense Colony*, KRL Road, Rawalpindi: Sciences Publishers.

Bohm, D. 1952. "A Suggested Interpretation of Quantum The oryin terms of 'Hidden' Variables, and II." *Physical Review* 85: 166-193.

Boissinot, S., Alvarez, L., Giraldo-ramirez, J., and Tollis, M. 2014. "Neutral Nuclear Variation in Baboons (genus Papio) provides Insights into Their evolutionary and Demographic Histories." *American Journal of Physical Anthropology,* December, 155(4): 621-634.

Bower, B. 2017. "First Settlers Reached Americas 130,000 Years Ago, Study Claims." *Science News* 4/26/2017.

Brannen, P. 2018. *The ends of the World: Volcanic Apocalypses, lethal oceans, and Our Quest to Understand Earth's Past Mass Extinctions.* NY: HarperCollins Publishers.

Britten, R. (2010). "Transposable element insertions have strongly affected human evolution." *Proceeding of the National Academy of Science of the United States of America* 107: 19945-19948.

Campbell, J. 1974. *The Mythic Image*. Princeton, NJ: Princeton University Press.

Carey, N. 2013. *The Epigenetic Revolution: How Modern Biology is Rewriting Our Understanding of Genetics, Disease, and Inheritance.* New York: Columbia University Press.

Carroll, S. 2019. *Something Deeply Hidden: Quantum Worlds and the Emergence of Spacetime*. NY: Random House.

Chabre, Y. and Roy, R. 2009. "The Chemist's Way to Prepare Multivalency." In, *The Sugar Code: Fundamentals of Glycosciences*, ed., H. Gabius, Weinheim, Germany: Wiley-Blackwell, pages 53-70.

Chamovitz, D. 2012. *What A Plant Knows: A Field Guide to the senses*. NY, Scientific American.

Chang, I., He, M., and Lam. C. 2018. "Congenital Disorders of Glycosylation." *Annals of Translational Medicine* Dec. 6(24).

Cheng, V., Lau, S., Woo, P., and Yuen, K. "Severe Acute Respiratory Syndrome Coronavirus as an Agent Emerging and Reemerging." *Clinical Microbiological Review*, October 20(4): 660-694.

Chopra, D., Penrose, R., and Carter, B. 2017. *How Consciousness Became the Universe: Quantum Physics, Cosmology, Neuroscience, Parallel Universes*. Defense Colony, KRL Road, Rawalpindi: Sciences Publishers.

Christensen, W. 2017. "Does the Universe have Cosmological Memory? Does This Imply Cosmic Consciousness?" In, *How Consciousness Became the universe: Quantum Physics, Cosmology, Neuroscience, Parallel universes. Defense Colony, KRL Road, Rawalpindi: Sciences Publishers.* Claren, A. and Michie, D. "An effect of the uterine environment upon skeletal morphology in the mouse." *Nature* 181: 1147-1148.

Collins, a. 2018. *The Cygnus Key: The Denisovans, Gobekli Tepe, and the Birth of Egypt*. Rochester, VT: Bear & Company.

Conway Morris, S. 1998. "Early Metazoan Evolution: Reconciling Paleontology and Molecular Biology." *American Zoologist* 38: 86-877.

Craig, N, Caraigie, R. Gellert, M., and Lambowitz, A. 2002. *Mobile DNA II*. Washington, D.C., American Society for Microbiology Press.

Crow, J. 1997. "The High Spontaneous Mutation Rate: Is It a Health risk?" *Proceedings of the National Academy of Science*, 94: 8380-8386.

Davidson, E. 2006. *The Regulatory Genome: Gene Regulatory Networks in Development and Evolution*. Burlington, VT: Elsevier.

Davidson, E. and Erwin, D. 2010. "An Integrated View of Precambrian Eumetazoan Evolution." *Cold Spring Harbor Symposia on Quantitative Biology* 74: 1-16.

Darwin, C. 1988 (orig. 1859). *On the Origin of Species*, 6th Edition. NY: New York University Press.

Darwin, C. 2019 (orig. 1871). *The Descent of Man.* Overland Park, KS: Digireads.com Publishing.

Dawkins, R. 1996. *Climbing Mount Improbable*. NY: Norton.

Dawkins, R. 1986. *The Blind watchmaker.* NY: W. WE. Norton & Company.

Dawkins, R. 1999. *The Extended Phenotype*. NY: Oxford University Press.

Dawkins, R. 2006 (orig. 1976). *The Selfish Gene*. NY: Oxford University Press.

Dawkins, R. and Wong, Y. 2017. *The Ancestor's Tale*. London, UK: Weidenfeld & Nicolson.

Deloison, Y. 2004. "A New Hypothesis on the Origin of Hominoid Locomotion." In, *From Biped to Strider: The emergence of Modern Human Walking, Running, and Resource Transport*, eds.

D. Meldrum and C. Hilton, pages 35-48, NY: Kluwer Academic/ Plenum Publishers.

DeWitt, B and Graham, N. (eds.) 2015. *The Many Worlds Interpretation of Quantum Mechanics.* Princeton, NY: Princeton University Press.

Dorey, F. 2019. "Sahelanthropus tchadensis." *Australian Museum,* 7/1/2019.

Douglas, A. 2010. *The Symbiotic Habit.* Princeton, NJ: Princeton University Press.

Dunning Hotopp, J., Clark, M., Oliveira, D., Foster, J., Fischer, P., Munoz Torres, M., Kumar, N., Ishmael, N., wang, S., Ingram, J., Nene, R., Shepard, J., Yomkins, J., Richards, S., Spiro, D., Ghedin, E. Slatko, B., Tettelin, H., and Werren, J. 2007. "Widespread lateral Gene transfer from Intracellular Bacteria to Multicellular Eukaryotes." *Science* 312, September, Issue 5845: 1753-1756.

Dupressoir, A., Lavialle, C. and Heidmann, T. (2012)."From Ancestral Infectious Retroviruses to Bona Fide Cellular Genes: Role of the Captured Syncytins in Placentation." *Placenta* 33: 663-671.

Durrett, R. and Schmidt, D. 2008. "Waiting for Two Mutations: With Applications to Regulatory Sequence Evolution and the Limits of Darwinian Evolution." *Genetics* 180: 1501-11509.

Eberlin, M. 2019. *Forethought: How the Chemistry of Life Reveals Planning and purpose.* Seattle, WA: Discovery Institute Press.

Edelman, G. 1993. *Topobiology: An Introduction to Molecular Embryology.* NY: Basic Books.

Eden, M. 1997. "Inadequacies of Neo-Darwinian Evolution as a Scientific Theory." In, *Mathematical Challenges to the Neo-Darwinian Interpretation of Evolution,* ed., P. Moorhead and M. Kaplan, Wistar Institute Symposium Monograph, No. 5, NY: Liss.

Erwin, D. Valentine, J. and Sepkoski. 1987. "A Comparative Study of Diversification Events: The early Paleozoic Versus the Mesozoic," *Evolution* 41: 1177-1186.

Everett, H. 1957. "Relative State Formulation of Quantum Mechanics." In, *Quantum Theory and Measurement*, J. Wheeler and W. Zurek (eds.), Princeton, NJ: Princeton University Press.

Festinger, L. 1956. *When Prophecy fails: A Social and Psychological Study of a Modern Group that Predicted the Destruction of the World*. NY: Harper-Torchbooks.

Faria, V and Sucena, E. "Novel Endosymbioses as a Catalyst of Fast Speciation." In, N. Gontier (ed.), *Reticulate Evolution: Symbiogenesis, Lateral Gene Transfer, Hybridization and Infectious Heredity*. NY: Springer, pages 107-120.

Farnsworth, K., Ellis, G., and Jaeger, L. 2017. "Living through Downward Causation: From Molecules to ecosystems." In, *From Matter to Life: Information and Causality*, eds. S. Walker, P. Davies, and G. Ellis, pages 301-333. NY: Cambridge University Press.

Fisher, R. 1918. "The Correlation Between Relatives on the Supposition of Mendelian Inheritance". *Transactions of the Royal Society of Edinburgh* 52 (2): 399–433.

Fodor, J. and Piattelli-Palmarini, M. 2011. *What Darwin Got Wrong*. NY: Farrar, Straus and Giroux.

Francis, R. 2011. *Epigenetics: How Environment Shapes Our Genes*. New York: W.W. Norton.

Fry, I. 2000. *The emergence of Life on earth: A Historical and Scientific Overview*. New Brunswick, NJ: Rutgers University Press.

Furey, J. and Fortunato, V. 2017. "The Theory of Mind Time." In, *Consciousness Became the Universe: Quantum Physics, Cosmology, Neuroscience, Parallel Universes*, 2nd Edition, pgs. 286-308. Science Publishers.

Fuss, J., Spassov, N., Begun, D., and Böhme, M. 2017. "Potential Hominin Affinities of Graecopithecus from the Late Miocene." *PLOS ONE* May 22: 12(5).

Gabius, H. 2000. "Biological Information Transfer Beyond the Genetic Code: The Sugar Code." *Naturwissenschaften* 87: 108-121.

Gabius, H. ed. 2009. *The Sugar Code: Fundamentals of Glyco-sciences.* Weinheim, Germany: Wiley-Blackwell.

Gabius, H. and Gabius, S. Eds., *Glycosciences: Status and Perspectives.* London: Chapman & Hall.

Gabius, H. 2018. "The Sugar Code: Why Glycans are So Important." *Biosystems* 164, February, pages 102-111.

Gabius, H. and Roth, J. 2017. "An Introduction to the Sugar Code." *Histochemistry and Cell Biology* 147: 111-117.

Galdieri, L., Mehrotra, S., Yu, S., and Vancura, A. 2010. "Transcriptional Regulation in Yeast During Diauxic Shift and Stationary Phase." *OMICS: A Journal of Integrative Biology*, Vol. 14, No. 6., pages 615-710.

Gao, S. 2017. "A Quantum Physical Effect of Consciousness." In, *How Consciousness Became the Universe: Quantum Physics, Cosmology, Neuroscience, Parallel Universes*, pages 509-517. Defense Colony, KRL Road, Rawalpindi: Sciences Publishers.

Garcia-Arocena, D. 2015. "The Genetics of Violent Behavior." *Blog on Genetics*, December 7, Bar Harbor, Maine: The Jackson Laboratory.

Gardner, A. and Conlon, J. 2013. Cosmological Natural Selection and the Purpose of the Universe." *Complexity* 19: 48-56.

Ghirardi, G. "Collapse Theories." In, *The Stanford Encyclopedia of Philosophy*, (Fall 2008 edition). E. Zalta (ed.), http://plato.stanford.edu/archives/fall2008/entries/qm-collapse/.

Gishlick, A., Matzke, N., and Elsberry, W. 2004. Meyer's Hopeless Monster. *Talk Reason.org*, September 12: http://www.talkreason.org/articles/meyer.cfm.

Gierlinski, G., Niedzwiedski, G., Lockley, M., Athanassiou, A., Fassoulas, C., Dubicka, Z., Boczarowski, A., Bennett, M. and Ahlberg, P. 2017. "Possible Hominin Footprints from the Late Miocene." *Proceedings of the Geologists' Association* 128: pages 697-710.

Giles, K., Woolnough, J., and Atwood, B. 2016. "ncRNA Function in Chromatin Organization." In, *Epigenetic Gene expression and Regulation*, pages 117-148, eds. S. Huang, M. Litt, and C. Blakey, NY: Academic Press.

Gillings, M. 2016. "Lateral Gene Transfer, Bacterial Genome evolution, and the Anthropocene." *Annals of the New York Academy of Science*, October 5.

Gleadow, A. 1980. "Fission Track age of the KBS Tuff and associated Hominid Remans in Northern Kenya." *Nature* 284: 225-230.

Goldschmidt, R. 1940. *The Material Basis of Evolution*. New Haven, CT: Yale University Press.

Gontier, N. 2015. "Historical and epistemological Perspectives on What Horizontal Gene Transfer Mechanisms Contribute to Our Understanding of Evolution." In, N. Gontier (ed.), *Reticulate Evolution: Symbiogenesis, Lateral Gene Transfer, Hybridization and Infectious Heredity*. NY: Springer, pages 121-178.

Gontier, N. (ed.). 2015. *Reticulate Evolution: Symbiogenesis, Lateral Gene Transfer, Hybridization and Infectious Heredity*. NY: Springer.

Gorkin, D., Leung, D., and Ren, B. (2014). "The 3D genome in transcriptional regulation and pluripotency." *Cell Stem Cell* 14:762-775.

Gould, S. and Eldridge, N. 1977. "Punctuated Equilibria: The Tempo and Mode of Evolution Reconsidered." *Paleobiology* 3: 115-151.

Gould, S. and Eldridge, N. 1993. "Punctuated Equilibrium Comes of Age." *Nature* 366: 223-227.

Hafer, A. 2015. *The Not-So-Intelligent Designer: Why Evolution Explains the Human Body and Intelligent Design Does Not*. Eugene, Or: Cascade Books.

Haig, D. 2012. "Retroviruses and the Placenta." *Current Biology*, Vol. 22, Issue 15.

Haldane, J. 1957. "The Cost of Natural Selection." *Genetics* 55:511-524.

Halfmann, R. and Lindquist, S.2010. "Epigenetics in the Extreme: Prions and the Inheritance of Environmentally Acquired Traits." *Science* 330, 623-630.

Hammeroff, S. 1998. "Funda-Mentality: Is the conscious mind subtly linked to a basic level of the universe?" *Trends in Cognitive Sciences* 2(4): 119-127.

Hammeroff, S. 2007. "Consciousness, neurobiology and quantum mechanics: The case for a connection." In, *The emerging Physics of Consciousness*, ed. J. Tuszynski, NY: Springer-Verlag.

Hammeroff, S. and Penrose, R. 1996. "Conscious events as orchestrated space-time selections." *Journal of Consciousness Studies*, 3(1), 36-53.

Hawking, S. and Mlodinow, L. 2012. *The grand Design*. NY: Bantam Books.

Hawks, J. 2009. "The Ardipithecus Pelvis." *John Hawks Weblog*, October 6.

Heijmans, B., Tobi, E., Stein, A., Putter, H., Blauw, G., Susser. E., and Lumey, L. 2008. "Persistent Epigenetic Differences Associated with Parental Exposure to Famine in Human." *Proceedings of the National Academy of Sciences of the United States of America*, 105: 17046-17049.

Hines, M. 2011. "Prenatal endocrine influences on sexual orientation and on sexually differentiated childhood behavior." *Frontiers in Neuroendocrinology*, Vol. 32, Issue 2 April, pages 170-182.

Huang, S., Litt, M., and Blakey, A. (eds.). 2016. *Epigenetic Gene Expression and Regulation*. New York: Academic Press.

Hubble, E. 1920. *Photographic investigations of faint nebulae*. Chicago, Ill: The University of Chicago Press.

Hitchens, C., Dawkins, R., Harris, S., and Dennett, D. 2019. *The Four Horsemen: The Conversation that Sparked an Atheist Revolution*. NY: Random House.

Hoffman, D. 2019. *The Case against Reality: Why evolution hid the Truth from Our Eyes*. NY: W. W. Norton & Company.

Huxley, T. 2012 (orig. 1863). *Man's Place in Nature and other Essays.* Amazon.com Services LLC.

Jansen, F. 2017. "The Observer's Now, Past and Future in Physics from a Psycho-Biological Perspective." In, *How Consciousness Became the Universe: Quantum Physics, Cosmology, Neuroscience, Parallel Universes*. Defense Colony, KRL Road, Rawalpindi: Sciences Publishers, pages 379-396.

Johns, T. 1999. "The Chemical Ecology of Human Ingestive Be haviors." *Annual Reviews of Anthropology* Vol. 28: 27-50.

Jorgensen, R. 2004. "Restructuring the genome in response to adaptive challenge: McClintock's bold conjecture revisited." *Cold Spring Symposia on Quantitative Biology* 69: 349-354.

Joseph, R. 2017a. "The Time Machine of Consciousness. Past, Present, Future Exist Simultaneously. Entanglement, Tachyons, relative Time, Circle of time, Quantum time, Dream time, PreCognition, Retrocausation, Déjà vu, and Premonitions." In, *How Consciousness Became the Universe: Quantum Physics, Cosmology, Neuroscience, Parallel Universes*. Defense Colony, KRL Road, Rawalpindi: Sciences Publishers, pages 309-315.

Joseph, R. 2017b. "The quantum Physics of God and how Consciousness Became the Universe." In, *How Consciousness Became the Universe: Quantum Physics, Cosmology, Neuroscience, Parallel Universes,* pages 543-572. Defense Colony, KRL Road, Rawalpindi: Sciences Publishers.

Kafatos, M., Tanzi, R., Chopra, D., Jones, F., Kennedy, J., and Kennedy, R. 2017. "How Consciousness Becomes the Physical universe." In, *Consciousness Became the Universe: Quantum Physics, Cosmology, Neuroscience, Parallel Universes*, 2nd Edition, pgs. 6-15. Defense Colony, KRL Road, Rawalpindi: Science Publishers.

Kak, S., Chopra, D., and Kafatos, M. 2017. "Perceived Reality, Quantum Mechanics, and Consciousness." In, *Consciousness Became the Universe: Quantum Physics, Cosmology, Neuro-science, Parallel*

Universes, 2nd Edition, pgs. 16-31. Defense Colony, KRL Road, Rawalpindi: Science Publishers.

Karanth, K. 2008. "Primate Numts and Reticulate evolution of capped and Golden Leaf Monkeys (Primates, Colobine). *Journal of Bioscience* 33:761-770.

Karanth, K. 2010. "Molecular Systematics and Conservation of the Langurs and leaf Monkeys of south Asia." *Journal of Genetics* 89: 393-399.

Karanth, K., Singh, L. Collura, R., and Stewart, C. 208). "Molecular Phylogeny and Biogeography of Langurs and Leaf Monkeys of South Asia (Primates, Colobine)." *Molecular Phylogenetics and Evolution* 46: 683-694.

Kauffman, S. 2019. *A World Beyond Physics: The Emergence and Evolution of Life*. NY, Oxford University Press.

Keller, C., Roos, C., Groeneveld, L., Fischer, J., and Zinner, D. 2010. "Introgressive Hybridization in Southern African Baboons Shapes Patterns of mtDNA variation." *American Journal of Physical Anthropology* May, 142(1): 125-136.

Khamsi, R. 2007. "Pubic Lice Leapt from Gorilla to Early Humans." *New Scientist*, March 7.

King, C. 1996. "Quantum mechanics, chaos and the conscious brain." *Journal of Mind and Behavior*, 18: 155-170.

King, C. 2017a. "Space, Time, and Consciousness." In, *Consciousness Became the Universe: Quantum Physics, Cosmology, Neuroscience, Parallel Universes*, 2nd Edition, pgs. 42-70. Defense Colony, KRL Road, Rawalpindi: Science Publishers.

King. C. 2017b. "Cosmological Foundations of Consciousness." In, *Consciousness Became the Universe: Quantum Physics, Cosmology, Neuroscience, Parallel Universes*, 2nd Edition, pgs. 86-99. Defense Colony, KRL Road, Rawalpindi: Science Publishers.

Koonin, E. 2012. *The Logic of Chance: The Nature and Origin of Biological evolution*. Upper Saddle River, NJ: Pearson Education.

Koonin, E. (2014). "Calorie restriction a Lamarck." *Cell* 158: 237- 238.

Krakauer, D. 2017. "Cryptographic Nature." In, *From Matter to Life: Information and Causality*, eds. S. Walker, P. Davies, and G. Ellis, pages 157-173. NY: Cambridge University Press.

Krakauer, D. (ed.). 2019. *Worlds Hidden in Plain Sight: The evolving Ideas of Complexity at the Santa Fe Institute*, 1984-2019. Santa Fe, NM: Santa Fe Institute.

Kuhn, T. 2012 (orig. 1962). *The Structure of Scientific Revolutions*. Chicago, Il: University of Chicago Press.

Laine, R. 1997. "The information-storing potential of the sugar code." In, H. Gabius and S. Gabius. Eds., *Glycosciences: Status and Perspectives*. Chapman & Hall, London, pp 1–14.

Laland, k., Uller, T., Feldman, M., Sterelny, K., Miller, G., Moczek, Jablonka, E., and Oding-Smee. 2015. "The extended Evolutionary Synthesis: Its Structure, Assumptions and Predictions." *Proceedings of the Royal Society*, August 22.

Lane, N. 2015. *The Vital question: Energy, evolution, and the Origins of Complex life*. NY: W. W. Norton.

Lederberg, E. 1950. "Lysogenicity in Escherichia coli strain K-12." *Microbial Bulletin* 1: 5-8.

Lederberg, E. 1953. "*Genetic* Studies of Lysogenicity in Escherichia Coli." *Genetics* 38 (91): 51-64.

Levinton, J. 1988. *Genetics, Paleontology, and Macroevolution*. Cambridge: Cambridge University Press.

Leisola, M. Witt, J. 2018. Heretic: One Scientist's Journey From Darwin to Design. Seattle, WA: Discovery Institute Press. Li, Z., Rosenbaum, M., Venkataraman, A., Tam, T., Katz, E., and Angenent, L. 2011. "Bacteria-Based AND Logic Gate: A Decision-Making and Self-Powered Biosensor." *Chemical Communications*, 47: 3060-3062.

Libet, G., Wright, E., and Pearl, D. 1983. "Time of conscious intention to act in relation to onset of cerebral activity (readiness-potential)

— The unconscious initiation of a freely voluntary act." *Brain* 106: 623-642.

London, F. and Bauer, E. 1983 (orig. 1939). "The theory of observation in quantum mechanics." In, *Quantum Theory and Measurement*, ed. J. Wheeler and W. Zurek, Princeton, NJ: Princeton University Press.

Lu, N., Wardell, S., Burnstein, K., Defranco, D., Fuller, P., Giguere, V., Hochberg, R., McKay, L., Renoir, J., Weigel, N., Wilson, E., McDonnell, D., and Cidlowski, J. (2006). "The pharmacology and classification of the nuclear receptor superfamily: glucocorticoid, mineralocorticoid, progesterone, and androgen receptors". *Pharmacological Review* 58 (4): 782–797.

Lupski, J. (2013). "Genetics: genome mosaicism – one human, multiple genomes." *Science* 341:358-359.

Lynch, M. 2006. "The Origins of Eukaryotic Gene Structure." *Molecular Biology and Evolution* 23: 450-468.

Mallo, M., Wellik, D., and Deschamps, J. (2010). "*Hox* genes and regional patterning of the vertebrate body plan." *Developmental Biology* 344: 7-15.

Margulis, L. 1999. *Symbiotic Planet: A New Look at Evolution*. NY: Basic Books.

Margulis, L. and Sagan, D. 2002. *Acquiring Genomes: A Theory of the Origins of Species*. NY: Basic Books.

Marks, R., Behe, M., Dembski, W., Gordon, B., and Sanford, J. (eds.). 2013. *Biological Information: New Perspectives*. Hackensack, NJ: World Scientific Publishing.

Marshall, P. 2015. Evolution 2.0: *Breaking the Deadlock Between Darwin and Design*. Dallas, T: BenBella Books.

Martin, F., Carminati, F., and Carminati, G. 2017. "Synchronicity, Entanglement, Quantum Information and the Psyche." In, *How Consciousness Became the Universe: Quantum Physics, Cosmology, Neuroscience, Parallel Universes*, pages 397-405. Defense Colony, KRL Road, Rawalpindi: Sciences Publishers.

Martincorena, I., Seshasayee, A., and Luscombe, N. 2012. "Evidence of Non-Random Mutation Rates Suggests an Evolutionary Risk Management Strategy." *Nature* 485: 95-98.

McBreaty, S. and Jablonski, N. 2005. "First Fossil Chimpanzee." Nature 437, pages 105-108. McGinnis, W. and Kurziora, M. 1994. "The Molecular Architects of Body Design." *Scientific American* 270: 58-66.

Mensky, M. 2017. "Logic of quantum Mechanics, Parallel worlds and Phenomenon of Consciousness." In, *How Consciousness Became the Universe: Quantum Physics, Cosmology, Neuroscience, Parallel Universes*, Pages 482-493. Defense Colony, KRL Road, Rawalpindi: Sciences Publishers.

Mesgaran, M., Lewis, M., Ades, P., Donohue, K., Ohadi, S., Li, C., and Cousens, R. 2016. "Hybridization can Facilitate Species Invasions, even without Enhancing Local Adaptation." *Proceedings of the National Academy of Sciences*, September 6.

Meyer, S. 2009. *Signature in the Cell: DNA and the evidence for Intelligent Design*. NY: HarperOne.

Meyer, S. 2013. *Darwin's Doubt: The Explosive Origin of Animal Life and the Case for Intelligent Design*. NY: HarperOne.

Mill, B, McBride, C., and Riddle, N. 2016. "Epigenetic Inheritance." In, *Epigenetic Gene Expression and Regulation*, pages 183-208, eds. S. Huang, H. Litt, and C. Blakey, New York: Academic Press.

Miller, G. 1956. "The Magical Number Seven, Plus or Minus Two: Some Limits on Our Capacity for Processing Information." *Psychological Review* 63: 81-97.

Mitchell, C. and Silver, D. 2018. "Enhancing Our Brains: Genomic Mechanisms Underlying Cortical Evolution." *Seminars in Cell & Developmental Biology* 76, April, Pages 23-32.

Mitchell, M. 2009. *Complexity: A Guided Tour*. NY: Oxford University Press.

Mitchell, C., Hobcraft, J., McLanahan, S. et al. (2014). "Social disadvantage, genetic sensitivity, and children's telomere length." *Proceedings of the National Academy of Science of the United States of America* 111: 5944-4949.

Mulligan, C. 2016. "Early Environments, Stress, and the Epigenetics of Human Health." In, *Annual Review of Anthropology* 45, 2016.

Nafgel, T. 2012. *Mind and Cosmos: Why the Materialist Neo-Darwinian Conception of Nature is Almost Certainly False*. NY: Oxford University Press.

Newsmax Health. 2018. "Common Chemical Causes Language Delay in Children." November 23.

Noble, D. 2017. "Digital and Analogue Information in organisms." In, *From Matter to Life: Information and Causality*, eds. S. Walker, P. Davies, and G. Ellis, pages 114-129. NY: Cambridge University Press.

Ochman, H. Lawrence, J., and Groisman, E. 2000. "Lateral Gene Transfer and the Nature of Bacterial Innovation." *Nature* 405: 299-304.

Pal, C., Papp, B., and Lercher, M. 2005. "Adaptive Evolution of Bacterial metabolic Networks by Horizontal Gene Transfer." *Nature Genetics* 37 (12), December, pages 1372-1375.

Parrington, J. 2015. *The Deeper Genome*. New York: Oxford University Press.

Patsos. G. and Corfield, A. 2009. "O-Glycosylation: Structural Diversity and Functions." In, H. Gabius, ed., *The Sugar Code: Fundamentals of Glycosciences*, Weinheim, Germany: Wiley-Blackwell, pages 111-137.

Penrose, R. 1998. *The Emperor's New Mind*. NY: Oxford University Press.

Penrose, R. 1994. *Shadows of the Mind*. NY: Oxford University Press.

Penrose, R. and Hameroff, S. 2017. "Consciousness in the Universe: Neuroscience, Quantum Space-Time Geometry and Orch OR

Theory." In, *Consciousness Became the Universe: Quantum Physics, Cosmology, Neuroscience, Parallel Universes*, 2nd Edition, pgs. 240-284. Defense Colony, KRL Road, Rawalpindi: Science Publishers.

Perkins, T. and Swain, P. 2009. "Strategies for Cellular Decision-Making." *Molecular Systems Biology*, 9:153-224.

Peters, J. (2014). "The role of genomic imprinting in biology and disease: an expanding view." *Nature Reviews Genetics* 15: 517-530.

Pfennig, K., Kelly, A., and Pierce, A. 2016. "Hybridization as a Facilitator of Species range expansion." *Proceeding Biological Sciences*, September 28, 283(1839).

Platt, R. Vandewege, M., Kern, C. et al. (2014). "Large Numbers of Novel miRNAs Originate from DNA transposons and are coincident with a large species radiation in bats." *Molecular Biology and Evolution* 31: 1536-1545.

Provine, W. 1971. *The Origins of Theoretical Populations Genetics*. Chicago, Il: University of Chicago Press.

Qui, J. 2016. 'How China is Rewriting the Book on Human Origins: Fossil Finds in China are Challenging Ideas about the Evolution of Modern Humans and our Closest Relatives." *Nature*, July 12.

Rebollo, R., Romanish, M., and Mager, D. (2012). "Transposable Elements: an Abundant and Natural Source of Regulatory Sequences for Host Genes." *Annual Review of Genetics* 46: 21-42.

Reich, D. 2018. *Who We Are and How We Got Here: Ancient DNA and the Science of the Human Past*. NY: Pantheon Books.

Rhen, T. and Cidlowski, J. 2005. "Anti-inflammatory Action of Glucocorticoids – New Mechanisms for Old Drugs". *New England Journal of Medicine*, 353 (16), October: 1711–23.

Ricard, P. and Hood, W. 2020. "Here's What Scientists Do and Don't Know About Wuhan Coronavirus So Far." *Science Alert*, January 29.

Richmond, B., Begun, D., and Strait, D. (2001). "Origin of human bipedalism: The Knuckle-walking Hypothesis is Revisited." *American Journal of Physical Anthropology, Supplement* 33: 70–105.

Riley, M., Faulkner, G., Dubnau, J., et al. (2013). "The role of transposable elements in health and diseases of the central nervous system." *The Journal of Neuroscience* 33: 17577-17586.

Rocas, A. and Carroll, S. 2006. "Bushes in the Tree of Life." *Plant Biology* 4, No. 11:1899-1904.

Roos, C., Zinner, D., Kubatko, L., Schwarz, C., Yang, M., and Meyer, D. (2011). "Nuclear versus Mitochondrial DNA: Evidence for Hybridization in colobine Monkeys." *BMC Evolutionary Biology* 11:77.

Roseboom, T., van der Meulen, J., Ravelli, A., et al. (2001). "Effects of prenatal exposure to the Dutch famine on adult disease in later life: an overview." *Molecular and Cellular Endocrinology* 185: 93-98.

Rosen, D. 1985. *Anticipatory Systems*. Oxford, UK: Pergamon Press.

Rossie, J., Ni, X., and Beard, C. 2006. "Cranial Remains of an Eocene Tarsier." *Proceeding of the National academy of Science*, March 21, 103(12) 4381-4385.

Rudan, I., Rudan, D., Campbell, H., Carothers, A., Wright, A., Smolej-Narancic, N., Janicijevic, B., Jin, L., Chakraborty, R., Deka, R., and Rudan, R. 2003. "Inbreeding and Risk of Late Onset Complex Disease." *Journal of Medical Genetics* 40: 925-932.

Rüdiger, G. and Gabius, H. 2009. "The Biochemical Basis and Coding Capacity of the Sugar Code." In, *The Sugar Code: Fundamentals of Glycosciences*, ed., H. Gabius, Weinheim, Germany: Wiley-Blackwell, pages 3-13.

Rüdiger, G. and Gabius, H. 1999. "Eukaryotic Glycosylation: Whim of Nature or Multipurpose Tool?" *Cell Molecular Life Science* 55:368–422.

Rüdiger H, Gabius H-J (2009) "The biochemical basis and coding capacity of the sugar code." In: H. Gabius, ed. 2009. *The Sugar Code. Fundamentals of Glycosciences*. NY: Wiley, Weinheim, pages 3–13.

Rupe, C. and Sanford, J. 2019. *Contested Bones*. NY: FMS Publications.

Rush, J. 1996. *Clinical Anthropology*. Westport, Conn.: Praeger Press.

Rush, J. 1999. *Stress and Emotional Health: Applications of Clinical Anthropology*. Westport, Conn.: Auburn House.

Rush, J. 2008. *Failed God: Fractured Myth in a Fragile World*. Berkeley, CA: North Atlantic Books.

Rush, J. 2011. *The Mushroom in Christian Art: The Identity of Jesus in the Development of Christianity*. Berkeley, CA: North Atlantic Books.

Rush, J. ed. 2013. *Entheogens and the Development of Culture*. Berkeley, CA: North Atlantic Books.

Ryan, F. 2009. *Virolution*. NY: Collins. Sagan, C. and Chyba, C. 1997. "The Early Faint Sun Paradox: Organic Shielding of Ultraviolet Labile Greenhouse Gasses." *Science* 276: 1217-1221.

Sanford, J. 2014. *Genetic entropy*. Longmeadow, MA: FMS Publications. Science News. 2019. "Human mutation rate has slowed recently." January 22, 2019.

Shams-Eldin, H., Debierre-Grockiego, F., and Schwarz, R. 2009. "Glycosylphosphatidylinositol Anchors: Structure, Biosynthesis and Functions." In, H. Gabius, ed, *The Sugar Code: Fundamentals of Glycosciences*, Weinheim, Germany: Wiley-Blackwell, pages 155-173.

Shapiro, J. 2011. *Evolution: A View from the 21st Century*. Upper Saddle River, NY:FT Press Science.

Shi, R. and Borgens, R. 1995. "Three-Dimensional Gradients of Voltage During Development of the Nervous System as Invisible Coordinates for the Establishment of Embryonic Pattern." *Developmental Dynamics* 202: 101-114.

Shubin, N. 2020. *Some Assembly Required: Decoding Four Billion Years of Life, from Ancient Fossils to DNA*. NY: Pantheon Books.

Sitchen, Z. 2014. *The Earth Chronicles*. Rochester, VT: Bear & Company.

Skinner, M., Gurerrero-Bosagna, C., Haque, M., Nilsson, E., Knop, J., Knutie, S., and Clayton, D. 2014. "Epigenetics and the Evolution of Darwin's Finches." *Genome Biology and Evolution* 6 (8): 1972-1989.

Snelling, A. 1998. "Anaesite Flows at Mt. Ngauruhoe, New Zealand, and the Implication for Potassium-Argon 'Dating'. *Presented at the Fourth International Conference on Creationism*, Pittsburgh, PA, August 3-8, 1998.

Sole, R. and Elena, S. 2019. *Viruses as Complex Adaptive Systems*. Princeton, NJ: Princeton University Press.

Stapp, H. 1993. *Mind, Matter, and Quantum Mechanics*. NY: Springer-Verlag.

Stapp, H. 2007. *Mindful Universe: Quantum Mechanics and the Participating Observer*. NY: Springer-Verlag.

Stapp, H. 2017. "Quantum Reality and Mind." In, *Consciousness Became the Universe: Quantum Physics, Cosmology, Neuroscience, Parallel Universes*, 2nd Edition, pgs. 32-41. Defense Colony, KRL Road, Rawalpindi: Science Publishers.

Stindl, R. (2014). "The Telomeric Sync Model of Speciation: Species-wide Telomere Erosion Triggers Cycles of Transposon-mediated Genomic Rearrangements, which Underlie the Saltatory Appearance of Non-adaptive Characters." *Die Naturwissencaften* 101: 163-186.

Stokstad, E. (2000). "Hominid Ancestors May have Knuckle Walked". *Science* 287 (5461): 2131–2132.

Syvanen, M. and Ducore, J. 2010. "Whole Genome Comparisons Reveals a Possible Chimeric Origin for a Major Metazoan assemblage." *Journal of Biological Systems* 18: 261-275.

Tammen, S., Friso, S., and Choi, S. (2013). "Epigenetics: the link between nature and nurture." *Molecular Aspects of Medicine* 34: 753-764.

Todesco, M., Pascual, M., Owens, G., Ostevik, K., Moyers, B., Hubner, S., Heredia, S., Hahn, M., Caseye, C., Bock, D., and Rieseberg, L. 2016. "Hybridization and extinction." *Evolutionary Applications*, August, 9(7): 892-908.

Trail. D., Watson, E., and Tailby, N. (2011). "The Oxidative State of Hadean Magmus and Implications for early Earth's Atmosphere." Nature, 480 (Dec): 79-82. *Universtoday*:https://www.universetoday.com/36302/atoms-in-the universe/#targetText=At%20this%20level%2C%20it%20is, hundred%20thousand%20quadrillion20vigintillion %20atoms.

Valentine, J. and Erwin, D. 1987. "Interpreting Great Developmental Experiments: The Fossil Record." In, *Development as an Evolutionary Process*, ed. R. Raff and E. Raff, NY: Wiley-Liss, 71-107.

Vannini, A. and Di Carpo, U. 2017. "Quantum Physics, Advanced Waves and Consciousness." In, *How Consciousness Became the Universe: Quantum Physics, Cosmology, Neuroscience, Parallel Universes*, pages 531-542. Defense Colony, KRL Road, Rawalpindi: Sciences Publishers.

Vass, R. 2017. "Multiverse Scenarios in Cosmology: Classification, Cause, Challenge, Controversy, and Criticism." In, *Consciousness Became the Universe: Quantum Physics, Cosmology, Neuroscience, Parallel Universes*, 2nd Edition, pgs. 428-442. Defense Colony, KRL Road, Rawalpindi: Science Publishers.

Von Newmann, J. 1932/1955. *Mathematical Foundations of Quantum Mechanics*. Princeton, NY: Princeton University Press.

Von Weizacker, C. 1980. *The Unity of Nature*. NY: Farrar Straus Giroux, Macmillan Publishers.

Wade, N. 2014. *A Troublesome Inheritance: Genes, Race, and Human History*. NY: The Penguin Press.

Walker, S., Davies, P., and Ellis, G. (eds.). 2017. *From Matter to Life: Information and Causality*. NY: Cambridge University Press.

Ward, P. 2018. *Lamarck's Revenge: How epigenetics is Revolutionizing Our Understanding of evolution's past and Present*. NY: Bloomsbury Publications.

Weikaert, R. 2004. *From Darwin to Hitler: Evolutionary ethics, Eugenics and Racism in Germany*. NY: Palgrave MacMillan.

Wells, J. 2001. "Making Sense of Biology: The Evidence for Development by Design." In, *Signs of Intelligence: Understanding Intelligent Design*, ed. J. Kushiner, and W. Dembski, Grand Rapids, MI: Brazos.

Weyrich, L. 2015. "Evolution of the Human Microbiome and Impacts on Human Health, Infections, Disease, and Hominid Evolution." In, N. Gontier (ed.), *Reticulate Evolution: Symbiogenesis, Lateral Gene Transfer, Hybridization and Infectious Heredity*. NY: Springer, pages 231-253.

Wigner, E. 1964. "Events, Laws of Nature, and Invariance Principles." *Science* 145(3636): 995-999.

Wigner, E. 1967. *Symmetries and Reflections*. London: Indiana University Press.

Wrangham, R. 2010. *Catching Fire: How Cooking Made Us Human*. NY: Basic Books.

Wright, Sewall (1984). *Evolution and the Genetics of Populations: Genetics and Biometric Foundations,* New Edition. University of Chicago Press.

Yockey, H. 2005. *Information Theory, Evolution, and the Origins of Life*. NY, Cambridge University Press.

Zenil, H., Schmidt, A., and Tegner, J. 2017. "Causality, Information, and Biological Computation: An Algorithmic Software Approach to Life, Disease, and the Immune System." In, *From Matter to Life: Information and Causality*, pages 244-279, eds. S. Walker, P. Davies, and G. Ellis. NY: Cambridge University Press.

Zhu, Z., Dennell, R., Wu, Y., Qui, S., Yang, S., Rao, Z., Hou, Y., Xie, J., Han, J., and Ouyang, T. 2018. "Hominin Occupation of the Chinese Loess Plateau since about 2.1 Million Years Ago." *Nature*, 559, pages 608-612.

Zimmer, C. 2015. *A Planet of Viruses*, 2nd Edition. Chicago, Il: University of Chicago Press.

Zorich, Z. 2010. "Ardipithecus: Ape or Ancestor?" *Archaeology Magazine* 63(1).

Zuber, C. and Roth, J. 2009. "N-Glycosylation." In, *The Sugar Code: Fundamentals of Glycosciences*, ed., H. Gabius, Weinheim, Germany: Wiley-Blackwell, pages 87-110.

Index

A

abortions – 67
acetabulum – 176
Acheulian hand axe – 143
Aegyptopithecus Zeuxis – 159
African Americans – 67
aging – 67-69, 97
Allah – 6, 125, 191-192
aluminum – 12
amygdala – 68
Aminata muscaria – 191
ancient aliens – 9, 11
arsenic – 55
archaea – 58, 60, 86, 89, 93, 96, 98, 102, 104-105, 109-110, 113
Argentina – 176
Austin, S. – 195
Australopithecus afarensis – 3, 22, 159, 161, 163, 171, 174, 176-177
Australopithecus africanus – 3, 167
autoimmune diseases – 65
autonomic nervous system – 84
Axe, D. – 195
Axolotls – 76

B

Bacteria – 2, 11, 15, 18, 21, 32, 54, 58, 60, 76, 77-80, 86-89, 93, 95-96, 98-110, 112-114, 184, 199
bacteriophage – 111
Bahinia – 164
balance – 12, 79, 96, 98-99, 119, 178
Barred tiger – 120
Salamander – 76, 120
Bateson, G. – 195
Bauhin, G. – 157
Bauhin, J. – 157
Behe, M. – 1-2, 20, 29, 31, 34, 36, 38, 43-44, 107, 112, 132, 140, 147, 149, 181
binomial nomenclature – 157
bipedalism – 22-23, 70, 123, 162-164, 168-169, 176-178, 186
birth control – 67, 81
black smokers - 17
"blank slate" – 85
blood clotting – vi, 1, 20, 29, 34, 36, 132, 187-188
blood pressure - 108

border towns – 121
Borneo – 157
Bose-Einstein condensation – 151
bosons – 186
bottlenecking – 117-118
boundary territories – 121
Brahma – 10, 126
Brahman – 126
Brannen, P. – 50
breeding population – 47, 166
Bristlecone pine tree – 175
Brussels Sprouts – 79
Bulgaria – 169
butyl benzyl phthalate (BBP) – 81

C

California – 55
California tiger salamander – 120, 173
Cambrian – vii, 1-2, 19, 75, 77-78, 104-105, 186
carbohydrates – 88, 90, 95, 98-100, 119
carbon – 11-14, 40, 76, 150, 159, 170, 172
carbon-14 dating – 22, 172
carbon cycle – 172
Caspian Sea – 167
cats – 6, 54, 57, 65, 77, 100-101, 111-112, 117, 120, 135
cell growth and proliferation – 92, 175
cells – vi, 1-2, 15, 17-21, 23, 27, 29-34, 36, 38-40, 45, 47, 54, 56, 58-64, 67, 69-76, 78, 80, 88, 90-94, 100, 102, 104-113, 118-119, 132-133, 135, 138-141, 153. 155, 160, 165-166, 174-175, 178, 180, 184-185
Chamovitz, D. – 113
chance – 39-40, 43, 46-47, 57, 68, 112, 118, 148-150, 167, 1845
Crick, F. – 58
China – 57, 107, 157, 173, 190
Christianity – 5-6, 125, 190-191
Chabre, Y. – 91
Chad man – 169
chimpanzee – 72, 74, 108, 122, 157, 159, 160-163, 166, 168-169, 174-180, 189
China – 57, 107, 157, 173, 190
chromosome – 54-55, 59, 65, 67-68, 72, 84, 102, 117, 121, 160
citric acid cycle – 1

clads – 74
clinical psychology – 100
codes – vi, 6-7, 15-16, 18, 20-22, 27, 30-35, 42, 53-62, 64-65, 67-70, 72-75, 77-79, 81, 83-93, 96, 101-102, 106-108, 111-114, 117-119, 123, 130-131, 140-141, 160, 178, 185-186, 188
coding sequence – vi, 20-22, 31-33, 5354, 56-59, 61-62, 6465, 6770, 73-74, 77-79, 81, 83,-88, 101108, 112-114, 117- 119, 130, 140-141, 160, 178, 186
computer models – 10, 18, 30, 168
Congruent Construction – 5, 185
Copenhagen interpretation – 154
Coronaviridae – 106
Coronavirus – 106-107
COVID-19 – 106-107
CpG mark – 53, 57
cranium – 124, 164, 167-169
crawling – 176, 178-179
Crete – 165, 177
Crocodiles - 96
Crow, J. – 46
Crow - 100

D

Darwin, C. – v, vii, 1, 4, 27- 29, 34, 47, 56, 71, 77, 150, 157, 159, 181, 183, 192-193
Dawkins, R. – vii, 4, 15, 27, 30-31, 33, 49, 73, 94, 151, 182, 187-188, 190-193
Debierre-Grockiego, F. – 92
de Bary, H. – 114
decision making – 19-20, 32, 45, 73-75, 92, 110, 115, 135-136, 141, 186
Deep Ocean Vent Model – 17
Deloison, Y. – 162-164, 173-174
Demodex brevis – 21
Demodex folliculorum – 21
demons – 2, 4, 6, 29, 32, 41, 125, 181, 184-185, 191
dendrochronology – 173
Denisovans – 21-22, 123, 158, 174
dental apes - 159
developmental plasticity – 37
Devil – 6
dextrorotatory – 15
dibutyl phthalate (DBP) – 81
Dinofelis barlowi – 167
Dmanisi – 166-167

DNA methyltransferase – 57
DNA – v, 2, 4, 7, 12, 15-16, 17, 29-30, 32-33, 38, 43, 53-63, 67-68, 71-72, 76-79, 81, 8486, 88, 90, 92,102, 105-106, 108, 110, 113, 155, 160, 178, 181, 185
Dopamine – 84
Doppler Shift – 7
Douglas, A. – 95
Dreaming – 95
Dutch Hunger Study – 54, 62, 94, 119, 186

E

Eberlin, M. – 30, 34, 38
Egyptian mummies – 172
Egyptian Plover bird – 96
Enceladus – 13
endocellular selection – 45, 64, 112, 166
endocytosis – 102, 110
endogenous antioxidants – 118
endosymbiotic hypothesis – 102
enterobacteria phage – 111
entropy – vii, 12, 19, 50, 54, 69, 74, 111, 118, 133, 183
Eocene – 164
epithelia – 64
epigenetic-influenced diseases – 62-63, 65, 72

epigenetics – 4, 18, 20, 25, 50, 53, 55-59, 61, 63-65, 67, 69, 71-77, 79-81, 83, 85, 87, 89, 91, 93, 112-113, 182
epiphyses – 178
estrogen – 81, 178
Eukaryotes – 20, 60, 76, 78, 87, 89, 102-103, 105, 109-110, 138, 150, 186
Eutheria – 21, 86, 109
evolutionary developmental biology (evo-devo) – 85, 175
exaptation – 164
Extended Evolutionary Synthesis – 36-37
extinction – 45, 47, 64, 71, 78, 108, 118, 163

F

face – 93, 124, 167
faith – 1-2, 4-5, 11, 14, 16, 20, 30-31, 40, 44, 77, 147, 149, 192
femur – 163, 165, 168, 175-179
feral cats – 65, 117
Festinger, L. – 41
Fleagle, J. – 163
fleas – 95-96
Flores, Indonesia – 3, 166, 174

INDEX

foramen magnum – 164, 168-169
frame shift – 75
free will – 38, 54, 139, 141, 153

G

galaxies – 7-9
Gardner, A. – 31
generic coding/rules – 11, 16, 18-19, 57, 151, 156, 180, 184-185
genomics – 22, 37, 85, 123, 166
genotype – 43, 58, 74
germ cells – 23, 47, 61, 67, 72, 175
Gibbon (Hylobates) – 167, 169, 180
Gierliński, G. – 165
Gilbert, S. – 175
Gillings, M. – 87, 112
Giraffa camelopardalis – 57
glenoid fossa – 175, 178
glucocorticoid receptor sites – 64-65, 68
glycans – 90, 92
glycolipids – 21, 90-91
glycolysis – vi, 1, 20, 29, 76, 88, 91
glycome – 88, 90
glycomic profile – 91
glycoproteins – 21, 90-92

God the Father – 6, 125, 191
Goliath – 158
gorilla (Gorilla beringei) – 108, 120, 122, 157, 159, 161, 163, 166, 169, 174, 180
Graecopithecus freybergi - 169
Great Barrier Reef – 57, 157
Guerrero, R. – 96
gut microbiome – 96-98, 100

H

heat shock – 92
Helicobacter pylori – 100
Hiliah (Allah) – 6, 125, 191-192
Hindu/Hinduism – 5, 126, 129, 131
Holy Mushroom – 191
Hominid – 161-162, 166-169, 171
Hominin – 3, 22, 73, 143, 157-159, 161, 163, 166-167, 169, 171, 173-175, 177, 179
Hominini – 158
Homo erectus – 3, 21, 123, 158, 166, 174
Homo ergaster – 158, 163, 167, 169-170
Homo floresiensis – 3, 166, 174
Homo habilis – 22, 123, 158
Homo neanderthalensis – 21,

158
Homo naledi
Homo rudolfensis
Homo sapiens
Honore, R. – 41
hormone – 81, 178
horseshoe crab - 187
Hound Dog Science – vii, 94, 142
HOX genes – 21, 70, 86
human leukocyte antigens (HLA) – 158
hydrochloric acid – 82, 99
hydrothermal vents – 18
hypocalcemia – 80
Hubble, E. – 7, 126
Humerus – 175-176, 178

I

Ialdabaoth – 125
iatrogenic illnesses – 81
ice ages – 4-5, 78
ilium – 168, 176-177
immune system/functions – 1, 21, 29, 64, 92, 102, 106, 108, 110, 113, 117, 119
imprinting – 59
inclusive inheritance – 37
information feedback – 140, 142, 175
informational systems – 42, 102, 113
intelligence – vii, 5-6, 10-12, 24, 30-31, 54, 78, 84, 129-130, 133, 142, 175, 183-184, 187, 193
irreducible complexity – 1, 18, 25, 29-30, 50, 77, 90, 132, 147, 181, 187
ischium – 176
Islam – 5-6, 125, 190-191

J

Jaglion – 120
Java – 3, 157
Jesus – 125, 191
Jim Crow laws – 190
Johns, T. – 118
Judaism – 5-6, 190-191
Jurassic – 176

K

Karanth, K. – 122
Kauffman, S. – 10, 43
Keller, C. – 117
Kimeu, K. – 169
knee joint – 70, 168, 175, 177-178, 180, 186
knuckle walking – 162, 176, 180
Kuhn, T. – 41, 60
Kyoto University – 17

L

Lane, N. – 17, 77, 104
Laine, R. – 88
Lakshmi – 126
λ (lambda phage, coliphage λ) – 111
lateral gene transfer – 86-87, 112,
lectin-carbohydrate interactions – 175
Leisola, M. – 38
Lemaître, G. – 7
leopard – 57, 167
Leptons – 186
lessor trochanter – 163
Levant – 80
levorotatory – 15
lice – 95, 120
Life Energy Potential (LEP) – 118
liberalism – 84, 188
Liger – 120
lightning – 17
Linnaeus, C. – 157
lipids – 12, 17-18, 21, 90-91
Litt, M. – 62,
Locke, J. – 84, 188
Lophs - 159
lotus flower – 126
Lovejoy, O. – 162, 168
lower esophageal sphincter – 82
Lucy – 3, 22, 159, 161, 163, 168-171, 176-177, 179
lysogeny – 110-11, 117
lytic/lytic enzymes – 110-111

M

Mammatus primigenius – 58
MAOA coding sequence – 84
Marx, K. – 23-24, 28, 182-184, 190, 192
Mastodon americanum – 58
mathematics 2, 7, 8, 10, 16, 30-31, 42, 49, 109, 112, 125, 133-135, 138, 142-143, 148, 160, 174, 181-184, 188
Maximum Life Span (MLSP) – 118
McClintock, B. – 62
MDR (minimal daily requirement) – 101
melanocytes – 80
memory – 84, 134, 138-139, 143. 155
Mesgaran, M. – 117
methane – 5, 13, 58
Meyer, S. – 30, 37, 39
Miocene – 56, 97, 101, 123, 158-159, 163, 173-174,

180
Mitchell, C. – 43, 68, 72
mites – 21
mitochondria – 54, 72, 75, 102, 105
modules – 29, 185
molecular clock – 160
mongoose – 96
Mono Lake, CA – 56
moon – 6, 10, 13-14, 17, 31, 49, 183
morality – 6, 130, 188-192
morphogenesis – 175
mRNA – 55, 60-61
Mussaurus patagonicus – 176
mutation – vi, 1-3, 5, 15, 19-21, 23, 27, 30-34, 36-38, 40-41, 43-47, 49-51, 55-58, 60, 62-64, 66, 70, 72, 74-76, 79-81, 86-88, 90, 93-94, 103, 106-109, 112, 118-119, 136, 140-141, 149, 160, 166-167, 178, 180-186, 188, 192
mutation-count – 45-46, 50

N

natural selection – vi, 2, 5, 20, 27, 31, 33-34, 36, 41, 43-45, 48-49, 51, 56-57, 64, 94, 112, 118, 150, 180-183, 185, 192
Nautilus pompilius – 57
Nautilus stenomphalus – 57
negative/positive heterosis – 117
neutron stars – 11, 172
niche construction – 37
nitrogen-14 – 172
NR3C1 coding sequence – 64, 66
nucleosome – 58
ncRNA – 60-61

O

observable universe – 39, 128, 147, 149
obesity – 59, 62, 66
octamer – 58
Okapia johnstoni – 57
Oligocene – 159
Orangutan (Pongo) – 157, 159, 161, 174
organosulfur thiourea – 79
Origin of Species – v, vii, 1-2, 4, 16, 20, 25, 28, 102, 111, 115, 183
Orrorin tugenensis – 3, 162-163
osteoclast cells – 178
ostriches – 96

P

Pacific Ocean – 17, 121
Pamir Mountains – 163
Pan paniscus – 122
Pan troglodytes – 122
Paranthropus – 161, 174, 176
parasitism – 95
Parmenides – 145
patrilocal – 166
Pfennig, K. – 117
phalanges – 162, 177-178
Phenacopithecus – 164
phenotype – 20, 32, 38, 53, 58, 60, 69, 74, 78, 108
phonemes – 141
phosphodiester bond – 53
phosphorus – 53, 55, 58, 104
phthalates – 180-181
phytochemicals – 96-97, 119
pig molars – 159
pin worms – 96
placenta – 21, 86, 107-109
Planck length – 40
Planck time – 40
plant/interspecies communication – 54, 87, 96, 113, 140
Pleiotropic – 84
politics – 4-5, 23, 41, 67, 127, 171
polydactylism – 117
Pompeii – 172
Potassium/Argon dating – 159, 170
preadaptations – 56
predestination – 43
protein folding – 34-35, 37-38, 92, 99
prototaxis – 103, 119
pseudo-scientists – 18
PTSD (Post Trauma Stress Disorder) – 68
PTC gene – 79-80
PTU – 79
pubis bone – 176

Q

quadrupedalism - 162
quantum decoherence – 128, 132
quantum biology – 25, 182
quantum mechanics – 127-128, 134, 136, 138, 151, 154-155
quarks – 186

R

Ran-Mut – 192
racemic – 15
red giants – 172
red shifting – 7

redundancy – 16, 34-35
redux reactions – 12
Reich, D. – 117, 123
retrotransposons – 62, 105
respiration – 102
reverse transcriptase – 110
rickets – 80
Rising Star Cave – 3
Rhodesian Man – 158
RNA – 4-7, 14-16, 18, 27, 29-30, 33-34, 37, 43, 51, 55-56, 60-61, 66, 70, 72, 78, 88, 90, 92, 94, 105, 110, 115, 140, 144, 146, 48-150, 152-153, 156, 178, 185, 188
Roos, C. – 122
Rossie, J. – 164
round worms – 95
Roy, R. – 91
Rudan, I. – 117
Rush, J. – 6, 42, 131, 143, 158, 189-191
Ryan, F.

S

Sahelanthropus tchadensis – 169, 174
salamander – 76, 120
Samotherium major – 56
sand dollar – 187
Sanger, M. – 67
SARS – 106-107
Sasaki, T. – 17
Saturn – 13
scapula – 178
Schwarz, R. – 92
sea cucumbers – 187
Second Law of Thermodynamics – 19, 138
self-organization – 10, 43, 85
Semnopithecus – 122
septicemia – 169
serotonin – 84
sexual signaling – 120
Shams-Eldin, H. – 92
shakti – 126
Shapiro, J. – 30, 54
Shubin, N. – 19, 36, 76, 184
silicone – 12-13
Silver, D. – 62, 72
singularity – 8, 155
Sitchen, Z. – 11
Sivapithecus indicus – 161, 174
sleep – 77-78, 82, 84, 120, 126, 162
Smith, J. – 45, 93
Snelling, A. – 170
Snowball Earth – 77-78, 104-105, 185-186

sodium – 64, 138
soma – 131
South Africa – 3, 157
Spencer, H. – v, 28
star fish – 187
statistics – 2, 37, 40-41, 132, 181
stress – vi, 19, 32, 45, 53-54, 56, 58-59, 62, 64-69, 73-75, 78-79, 82-84, 92, 94, 100, 103, 105-106, 114, 119, 121, 124, 158, 165-166, 175, 189
string theory – 9
sugar code – 20, 27, 88-93
Sumatra – 157, 174
symbiosis – 21, 72, 95-97, 103, 107, 112, 114
syncytin – 107-108
synergistic epistasis – 45-47, 50

T

Tajikistan – 163
Tajik National Park – 163
talk therapy – 68, 83, 100
Tanzania – 96
tapeworms – 95
Tarsier – 184
Tarsius eocaenus – 164
Taung infant – 3, 167
teeth/tooth – 96, 143, 159, 161-162, 169, 171
teleology – 10
telomeres – 67-68
temporal fascia – 161
Theria – 21, 109
third alphabet of life – 30, 88, 90, 93
thought/thinking - vii, 5,16, 19-20, 22, 25, 32, 34, 45, 54, 64, 66, 69, 73-76, 80, 106, 129, 135-136, 141, 143-144, 146, 151-152, 166, 180, 186
tibia – 175, 177-179
Todesco, M. – 117
topobiology – 19, 23, 25, 174-175, 182
totipotent cell – 133
Trachypithecus geei – 122
Trachypithecus pileatus – 122
triradiate cartilage – 177
Triassic – 176
tsunamis – 17

U

uniformitarianism – v
unintended design – 19, 184
upper esophageal sphincter – 82
USDA – 83, 97, 99-101, 119
UV radiation – 14, 17, 32, 140

V

velociraptor – 113
violence – 84
viruses – 18, 21-22, 62, 79-80, 96, 98-99, 102, 105-110, 113, 184
Vishnu – 126
von Weiszacker, C. – 142

W

Wade, N. – 85
Walker, A. – 169, 173
warthog – 96
warrior gene – 84
Watson, J. – 58
Weber, M. – 28
whipworms – 95
Whitehead, A. – 140
White, T. – 22, 162, 168, 171, 174
Wikipedia – 117, 120
Witt, J. – 38, 85-86, 108
Woolly Mammoth *(Mammatus primigenius)* – 58, 157
Wuhan, China – 107
WW II – 28, 66

Y

Yahweh – 125, 191-192
Y-5 cusp pattern – 159
Younger Dryas – 4

Z

zebra – 96, 120
Zhu, Z. – 158
Zonkey – 120

Author Profile

John A. Rush, Ph.D., N.D. (ret.) is Professor of Anthropology at Sierra College, Rocklin, Ca., and Folsom Lake College, Folsom, Ca. with specialties in human information processing, symbolism, and biological anthropology. Dr. Rush is also a retired Naturopathic Doctor, Clinical Anthropologist, and Medical Hypnotherapist in private practice from 1974 to 2008. Dr. Rush's publications include *Witchcraft and Sorcery: An Anthropological Perspective of the Occult* (1974), *The Way We Communicate* (1976), *Clinical Anthropology: An Application of Anthropological Concepts within Clinical Settings* (1996), *Stress and Emotional Health: Applications of Clinical Anthropology* (1999), *Spiritual Tattoo: A Cultural History of Tattooing, Piercing, Scarification, Branding, and Implants* (2005), *The Twelve Gates: A Spiritual Passage through the Egyptian Book of the Dead* (2007), *Failed God: Fractured Myth in a Fragile World* (2008), *The Mushroom in Christian Art: The Identity of Jesus in the Development of Christianity* (2011), and he is editor of and contributor to, *Entheogens and the Development of culture: The Anthropology and Neurobiology of Ecstatic Experience* (2013).

www.ingramcontent.com/pod-product-compliance
Lightning Source LLC
Chambersburg PA
CBHW071355210526
45465CB00001B/105